THE FINAL MYSTERY

DISCOVERING THE UNIMAGINABLE

NIGEL ALDWORTH

Copyright© 2014 Nigel Aldworth

All rights reserved.
No part of this publication may be reproduced, stored in a retrieval system, or transmitted in any form or by any means, without the prior permission in writing of the author, nor be otherwise circulated in any form other than that in which it is published and without a similar condition including this condition being imposed on the subsequent publisher.

ISBN-10: 149096035X
ISBN-13: 978-1490960357
TCODB001

To Lucy and Matt

Acknowledgments

I want to thank everyone who has helped in producing this book. They include all the well known scientists, many of them Nobel laureates, who have engaged with a broad audience and whose work I have quoted extensively. More particularly I want to thank Matthew Aldworth for his insightful contributions and endless encouragement, and Lucy Robinson for her enduring support and fixing my many typographic errors. Thanks also to Jac Büchner for editing the manuscript which I kept changing many times, and to Barbara Moore for proof reading. A big thank you must also go to John Stuart Wilson without whose belief and dogged persistence this book might not have seen the light of day.

Contents

Introduction i

Part One   Evolution 1
    Chapter One - In the Beginning 3
    Chapter Two - Life 9
    Chapter Three - Early Revolutions 21
    Chapter Four - The First Civilisation 28
    Chapter Five - Civilizations of the Middle East 38

Part Two   Discovering the Universe 49
    Chapter Six - Escaping from Ignorance 51
    Chapter Seven - The Universe Moves 66
    Chapter Eight - A Remarkable Person 72
    Chapter Nine - Evidence of Creation 87

Part Three   Exploring the Smallest 99
    Chapter Ten - The Atom 101
    Chapter Eleven - Crossing the Atomic Boundary 108
    Chapter Twelve - The Quantum Revolution 115
    Chapter Thirteen - The Age of Uncertainty 122
    Chapter Fourteen - Magic that's Real 129
    Chapter Fifteen - Infinite Paths 136
    Chapter Sixteen - The Great Smoky Dragon 143

Part Four   Where Worlds Meet 155
    Chapter Seventeen - Atomic Forces 157
    Chapter Eighteen - The Standard Model 167
    Chapter Nineteen - Creation from Nothing 178
    Chapter Twenty - The Multiverse 185

Part Five   The Biggest Mysteries 199
    Chapter Twenty One - The Mystery of Parallel Universes 201
    Chapter Twenty Two - The Mystery of No Time And No Space .. 210
    Chapter Twenty Three - The Mystery Beyond the Multiverse 219
    Chapter Twenty Four - Are There Limits To Reason? 238
    Chapter Twenty Five - The Final Mystery 255

Epilogue 270
Images and Illustrations 279
Bibliography 283
Index 291

Introduction

Something very big is happening within science. Something that is deeply disturbing. Although no one dares mention it, we may have arrived at a point where science can go no further. I don't mean the end of new technology or space missions, I mean the possibility of ever answering the really big questions like: is there a Theory of Everything and, as John Wheeler, a major figure of 20th century science once said: 'how come existence?' Increasingly, it seems, we may be staring at an ultimate abyss beyond which our way of knowing things can take us no further, where we are confronted with what is the final mystery – the origin of all information. That is what this book is about.

Although it sounds a bit daunting it is easy to understand. Ever since humans first became conscious they have wondered what caused everything to exist. So I want to go on this journey from the very beginning to the present moment. It is the story of humanity's search for an ultimate explanation of the universe. Leon Lederman, a famous Nobel Prize winner, once thought it would be a set of equations you could wear on your T-shirt. But recently science has discovered that it goes much deeper than this.

At the heart of science there lies a dark secret. One that is discussed in private but rarely mentioned in public. It is surrounded by an unspoken conspiracy of silence. We will be encountering some well-known scientists and mathematicians who have actually called it an 'underground religion', others a 'religious belief'. But over the years two scientists in particular have had the professional courage to speak about it openly. And they are not ordinary scientists. Both have substantial international reputations. The first is Sir Roger Penrose, the Emeritus Rouseball Professor of Mathematics at the University of Oxford, and the second is John Barrow, the Professor of Mathematical Sciences at the University of Cambridge. From this beginning many other prominent scientists like Marcus du Sautoy, David Deutsch, Max Tegmark and Peter Atkins are now discussing it openly. We will be investigating this mystery at the heart of science. Why? Because as we will see, it brings to light what must be the greatest revolution ever to confront humanity. It is a revolution that very few people realise is happening.

Currently, many scientists are worried about the issue of religion world wide. They have every reason to despair when polls in the U.S.A. consistently show that around 45% of U.S. citizens believe that God created humans within the last 10 000 years! Almost half the population does not believe in evolution. A poll conducted in the U.K. for the BBC in 2006 found that of 2 000 people questioned, only 48% believed in evolution. This is not just shocking, it's positively scary. No wonder scientists are concerned.

In a special issue New Scientist magazine reported that, "religious movements are sweeping the globe preaching unreason, intolerance and dogma."[1] What Richard Dawkins, a famous evolutionary biologist from the University of Oxford, likes to refer to as 'the primitive darkness that is coming back'. Unfortunately, it is no good telling people that religion is just wrong without giving them an alternative belief system. People need to believe in something. A three dimensional object with conscious awareness embedded in a world of cause and effect needs to know how everything happened. In the absence of an answer it will invent one and call it God. So powerful is this need, it can persuade people not only to defy logic and reason, and even the evidence of their senses, but in extreme cases to blow themselves up. Science *is* a belief system. One that is based on physical evidence, but it is not much good if only a few understand it and most find it boring.

With sectarian conflict becoming one of the defining issues of this century just as political ideology was of the 20th, there is now an urgent need to spread the word of science as an alternative belief system. That is my purpose in writing this book.

The truth is that there are far greater mysteries in science than there ever were in any religion. Nor is the road hard to follow. There is only one equation in the whole of this book, and there are many interesting and amusing anecdotes about the lives of the scientists involved. Right now

INTRODUCTION

with religious extremism on the increase, there is a pressing need for as many people as possible to know about the scientific journey.

I have chosen as our main guide a man who is widely regarded as the patriarch of modern theoretical physics and cosmology. His name is Steven Weinberg and he won a Nobel Prize in 1979 for uniting two of the forces that exist in Nature. He is a global figure of the scientific establishment. But I have chosen him also because he has a reputation as the scientists' scientist. He is well known for saying things like: "With or without religion, good people can behave well and bad people can do evil; but for good people to do evil – that takes religion."[2]

We will also meet many other famous scientists along the way. Because we are going to the very frontiers of knowledge, we will be meeting only the best that there is to guide us. I will introduce you to each of them as we make our way. Going back for a moment to earlier days, Steven Weinberg once said this:

> "The more the universe seems comprehensible, the more it also seems pointless...the effort to understand the universe is one of the very few things that lifts human life a little above the level of farce, and gives it some of the grace of tragedy."[3]

When I first read this, many years ago now, it took a while for it to sink in, but once it did it struck me as one of the most disturbing statements about our human condition ever written. Incidentally, it has often been quoted. Unfortunately this is what frightens people about science and makes them want to reject it or ignore it or belittle it. I want people to embrace it. Einstein once said that the greatest mystery in this universe is that it is capable of being understood. It's a statement that seems to crystallise the mystery of our existence.

Science has shown us that the human mind is the end product of the evolution of the entire universe, from the simplest structures created at the moment of the Big Bang, to the complex arrangement of nerve cells which is responsible for our consciousness. There is no 'magic' in between. So, besides the enormity of the cosmic scale of things, our barely visible smudge of consciousness cannot be unimportant. It represents the cutting edge of how our reality is evolving. This is surely the most astonishing destiny given to humankind, a destiny that carries with it an enormous responsibility.

It is time for us to grow up as a species and throw away the primitive crutches of religious and political bigotry. Time for us to finally rid our planet of disease and starvation. Time for us to end the mindless cruelty and violence of wars. We are confronting a destiny far greater than the need to kill each other over what is trivia by comparison.

The central odyssey of human consciousness has always been the quest to discover what caused everything to exist. This search for an ultimate explanation, whether it is a set of equations or something more fundamental, makes little difference. It is still, as Stephen Hawking once suggested, a search to understand the mind of God. What he called the final triumph of human reason. What we are going to discover is that we may never arrive at that moment what we are going to discover is the unimaginable – the origin of all information, which exists, but will forever remain incomprehensible to the human mind. We are going to discover the unknowable!

To achieve the best possible scientific authenticity I have restricted my sources to the published works of Nobel Laureates, Fields Medallists and professors from the top universities in the world. All quotations are referenced.

References

1. Editorial. 08 October 2005. *Reality wars*. New Scientist. No. 2520: p 39.

INTRODUCTION

2. Weinberg, Steven. 2003. *Facing up*. London: Harvard University Press. p 242.
3. Weinberg, Steven. 1987. *The First Three Minutes*. London: Fontana. p 149.

Part One

Evolution

Chapter One
In the Beginning

Before Space and Time – the basics

In the Beginning there was nothing…

No space and no time. There wasn't even darkness or light. This is what science calls the quantum vacuum. What does it look like?

Imagine you are in a plane flying high above an ocean. When you look out of the window the surface of the sea looks quite smooth and flat, but if you went down to a few hundred feet you would realise that it is not smooth at all, it is raging with waves and white foam. Or imagine a flat white beach. From a distance it looks flat and smooth, but if you were a tiny ant climbing over each grain of sand it would seem very different.

Imagine then, a great stretch of nothing. If you zoom in to the very tiniest level, much smaller than an atom, you would notice that it is not 'nothing' at all. It is made up of a seething turmoil of bits of atoms jumping in and out of existence very briefly. Physics calls them 'virtual particles' because they never become properly real. They can only exist for billionths of a second before disappearing back into nothing.

I should point out that this is not just a theory. The vacuum exists all around us today at the deepest level, and its presence can be measured in a fairly simple laboratory experiment known as the Casimir Effect.

The Big Bang

So, in the beginning there was just this vacuum of nothing. Suddenly, in as much as anything can be 'sudden' when there isn't any time, one of those virtual particles experienced an inappropriate delay in returning to the vacuum, what is called a 'fluctuation'. But it was too big to return to the vacuum:

KA-BOOM!

This tiny speck exploded. The first bit, the 'Ka' bit, is what scientists have called Inflation, a super massive, super fast expansion. Faster than anything ever since. The 'Boom' bit is what they call the Big Bang.

The biggest explosion ever to happen. But a very strange one. For a start, it was utterly silent. There was no one to hear it of course, nor was there a place where it could be seen. Space and time arrived with the explosion. Right here at the very beginning, we can see that space and time are derived from something else which will be important on our journey.

Out of this speck poured huge amounts of energy in the form of the first most primitive bits, like photons of light and electrons and quarks. In far less than the blink of an eye the enormous temperature began to drop as quarks joined up to form protons and neutrons. Shortly after, some of the electrons and protons got together to form the first atoms of hydrogen, with one electron and one proton each. Some of these then got together and formed the first helium atoms, and so the first two most simple types of atom were created.

All the stars we see in the night sky today are still made up of roughly 75% hydrogen and 25% helium. At this stage there was nothing to see yet because all the bits that ever existed were in a very tight space, banging together so ferociously that light was trapped.

After 380 000 years, space had become big enough for photons and other particles, that had not turned into hydrogen and helium, to escape. It is this radiation that space instruments like COBE and WMAP can measure to day. It is called the microwave background radiation. Surprisingly, when an old analogue TV set went on the blink and you just got a picture of black and white specks on the screen or noise, some of those specks you saw were actually part of this background radiation from the beginning of time – right there in your living room.

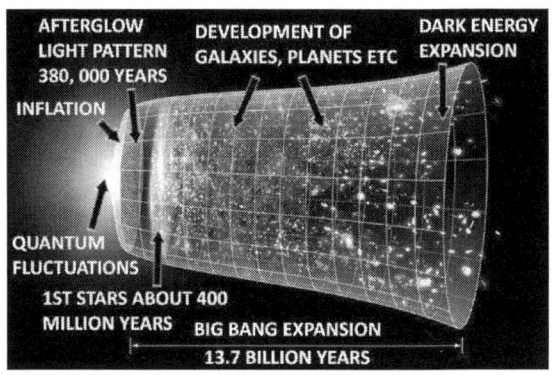

Figure 1 Inflation and Big Bang to the present

The period of Inflation was discovered by Alan Guth in 1981. It was of startling significance because, for the first time, science was able to explain how everything could be created out of nothing. Suddenly there was a real physical alternative to the Old Testament Genesis.

Not only was the theory consistent, it also answered some big problems about the universe that astronomers were having trouble with at the time. However, there is one rather disturbing side effect. It also predicts that there are other universes besides our own. Not just a handful either. Some versions of Inflation predict an infinite number of other universes.

Galaxies and Stars

Meanwhile, as space continued to expand, slight hiccups or bumps in the clouds of hydrogen and helium gas began to clump together under the attraction of gravity. It is thought that these clouds turned into the first mini galaxies called proto-galaxies.

Within these clouds local patches of hydrogen and helium atoms began to attract each other. Once this started they couldn't help being squeezed tighter and tighter together by gravity, until they were so compressed that nuclear fusion started to happen, as it does in our sun, and they began to ignite. So the first stars were born.

Like humans, stars have a life cycle. They are born, live out their lives and then die. If they didn't, we wouldn't be here.

As stars began to burst into life the first proper galaxies started to form. In the early universe galaxies were smaller and often collided with each other to make bigger ones. This is still going on. Galaxies are continually evolving like everything else in the universe. Our Milky Way is one of about thirty called the Local Group, which is small compared to other groups.

These groups of galaxies then form clusters, and clusters form super clusters, and super clusters form great chains, most of which are still assembling. In between the chains are vast voids that seem to be empty. Coincidentally, pictures of these chunks of universe look just like the dried lace of bubbles left on the inside of an empty beer glass. At the local level Andromeda, our nearest galactic neighbour, is streaming towards us at a frightening speed and will actually collide with the Milky Way in about two billion years.

Hopefully there will still be humans around to witness it, because the night sky will be an amazing spectacle of stars. They won't need to worry about being burned to a cinder. Stars are small compared to the space around them, so we are not likely to suffer any disruption to our little solar system. Unless we are pretty unlucky of course.

The Home Patch

So far the only things that exist in the universe are the two most primitive types of atom, along with lots of other smaller bits, like free electrons and protons, as well as photons of light and things called neutrinos. The reason that stars are so important is that they are engines for making more complicated atoms. As they burn out their lives they make the heavier, more complex atoms, like oxygen, nitrogen and iron.

Carbon is perhaps one of the most important and strangest atoms because it is what we are made out of. To make carbon requires a complicated dance between three naked helium atoms, (naked because they've lost their electrons). Then they have to do an elaborate jig at a terrible speed and, at a very precise instant when the forces are just right, pop! A carbon atom has been created. None of the other atoms require such an elaborate, almost 'contrived' looking dance. It seems *so* contrived you could be forgiven for thinking, as some people have, that it's a clever scheme from the mind of a grand designer, because without it there would be no intelligent life.

Our sun is a sort of medium size star, but there are obviously many that are bigger. In about another four billion years our sun will have burnt out most of its fuel and it will begin to die. In the process of dying it will expand into what astronomers call a red giant. It will grow so big it will swallow the Earth. Which will be the end of us if we are still here. Hopefully we'll have populated the rest of the galaxy by then. Any star that is slightly bigger than our sun has a different destiny. Instead of expanding outwards, it will shrink under the pull of gravity until it explodes in spectacular fashion, known as a supernova. Astronomers see them happening all the time, in other galaxies.

When a star explodes, it throws out into space all the heavier atoms it has made during its life and death, along with the remaining hydrogen and helium. All this debris forms clouds of gas and dust floating in space. One of the wonderful surprises of the Hubble telescope was that it managed to capture images of these huge clouds and within them a few infant stars being born.

So our sun is a second generation star that formed in this way from the leftovers of earlier stars containing all the 92 different atoms we know of. As it compressed under its own gravity and began to shine, it left lumps of rubble and gas orbiting around it. One of those bits of rubble we call Earth.

So we reside in the Milky Way galaxy which has a spiral shape, like a giant Catherine wheel, and on the outer edges of one of its spiral arms sits the sun with its planets for company. When you look up at the sky on a clear night you can see an arc of stars across the sky. What you are seeing is part of the spiral arm of our galaxy.

Distances

To give you a quick idea of the size of things, it takes eight minutes for light from the sun to reach us here, but five *years* for light to travel from the nearest star. Even with travelling at 186 000 miles (300 000 kms.) per second, yes seriously, per second, light from the other side of the Milky Way takes a hundred thousand years to get to us, so the light we see now, actually left when the humans first migrated from Africa using stone tools. And the light from Andromeda, which astronomers will see in their telescopes tonight, left there long before there was a human species.

As Stephen Hawking once remarked, we inhabit an insignificant planet orbiting a mediocre star in the outer suburbs of an ordinary galaxy, of which there are over a hundred billion in the observable universe. But, although at first sight it seems that humanity is utterly unimportant on the scale of the cosmos, the real truth as I said in the introduction, is that our barely visible smudge of consciousness is without doubt its single greatest achievement.

Chapter Two
Life

All that has ever happened

If you were to mention 'evolution' to most people they would think you were talking about Darwin and human evolution. But in fact human evolution is only a very tiny part of a much bigger thing that is going on all around us right now, and will continue into the future. In fact, all that has ever happened in the history of the universe is the story of ever increasing complexity, the evolution of 'matter' from the simplest structure of a pre-atomic plasma to the human brain. If there was ever an epic drama, this has to be it. And it looks as though we are playing a leading role. It almost seems as though whatever created the vacuum might have had us in mind.

From a cosmic perspective then, the only thing that has happened since time began is the evolution from the simplicity of nothing to the complexity that creates our consciousness. When we first opened our eyes, so to speak, the universe became aware of itself. Whether we like it or not, we are the children of the stars. And not only that, the Big Bang and the vacuum before it were actually pregnant with us and this 21st century moment in time. This is surely one of the most exciting realisations that science has given us.

The Early Earth

At first the earth was a hot molten ball of all the elements. As it cooled from the outside it formed a crust and the heaviest elements sank towards the centre. The earth's core is still molten iron. At the same time all the lighter elements bubbled their way to the surface and escaped from the crust through vast steaming vents and huge volcanoes to form a thick atmosphere that was trapped at the surface by the earth's gravity. It has been suggested that the oceans might have come from collisions with comets, but at this time the earth was still being bumped and bombarded by other bits of loose debris that hadn't yet formed into planets. The general consensus is that a particularly large collision broke a bit off the earth that clumped together to become our moon. The moon does not contain much iron, so it is thought to have been part of the earth's crust. The surface of the moon is full of impact craters, which confirms what a violent place the solar system must have been in those early days.

As the earth cooled, the atmosphere, which was mostly water vapour, began to condense so that it rained continuously for perhaps a million years or more, forming the oceans which now cover 70% of the earth's surface. Another step in the evolution towards greater complexity was taking place here. The water that formed the oceans is made out of two hydrogen atoms linking with one oxygen atom to make H_2O. In this way many of the rich variety of chemical compounds were being formed.

Once evolution has achieved chemical bonds, chains of these bonds can join up to form larger and larger chemical compounds like amino acids, (a bit like the galaxies are doing) which are the building blocks out of which DNA is made. And DNA is an essential ingredient of the next major evolutionary event. So far everything that evolution has managed to produce is dead, just lifeless chemicals. But another big leap is waiting off-stage.

Different Possibilities

The origin of life is still controversial because the transition from lifeless to living is difficult to cut precisely. In some ways there is no difference at all, or very little. It's just stuff slowly getting more complex and at some point we call it 'living'. So it seems to me that the question of how it came about is unimportant compared to the big picture. At the moment there are a few different scenarios as to how life came about.

One of these is that life came from space. When it was first proposed back in the 1960's by the astronomer Sir Fred Hoyle, people thought he was just mad. After all, he was a bit eccentric. He was a famous academic who wrote a science fiction novel about a cloud of cosmic gas that was intelligent. It was Sir Fred who gave the Big Bang its name in a BBC radio interview. He was making fun of it because the idea seemed too childishly simple. I'm inclined to agree about the name. It sounds like a child's party cracker. Why don't they call it something slightly more dignified, like: The Origin Event?

But anyway, he had the last laugh, because over the years it has been found that comets and asteroids do contain loads of organic substances, including hydrocarbons and amino acids. The most recent evidence comes from the 'Star Dust Probe' that brought back samples from the comet Wild 2. They included an abundance of organic compounds. In fact, there is now a whole profession that combines astronomy and biology. They are scientists called astrobiologists.

NASA faced a tricky problem when it was given the task of finding evidence of other life forms on Mars and on the moons of Jupiter and Saturn. Without a definition of life they wouldn't know what they were looking for. So they came up with "Life is a self-contained chemical system capable of undergoing Darwinian evolution."[1]

Another idea is that life might have arisen from silicon crystals because they are the most primitive things that can copy themselves. But currently the two most popular scenarios are that life began either in shallow lagoons or next to volcanic thermal vents deep under the oceans.

As I said, it's more difficult to define life than you think. Even continents and rocks and mountains evolve over geological time. And if you define it as something that breathes oxygen and eats fuel for energy, expels waste and has sex appeal, then you could be describing a Ferrari. The defining thing about life seems to be that it can reproduce itself in a way that allows for slight changes over succeeding generations. Viruses sit on the borderline between what's alive and what isn't. Some of them can't even reproduce by themselves and have to use other cells to do it in, which is what I guess makes us sick when they invade our bodies.

Perhaps the best way is to look back at the simplest organisms and see what we all have in common. Everything that's alive to day is made up of cells that contain just twenty amino acids along with the two nucleic acids, RNA and DNA. Scientists have so far managed to create RNA but not DNA, which is the key to life.

Most people have seen pictures of the double helix and how the two strands can twist together like a spiral staircase to form a structure that is probably unique in the whole universe because of the amount of information it can contain. If there is life on other planets, no matter how far away, then you could almost guarantee that it would have to be made of DNA. No other arrangement of matter on this tiny scale can carry so much information in such a small space.

It's like a sort of optimum structure that evolution contrived over millions of years. From just four bases labelled A, T, G and C it provides a 'blue-print' for every living thing. It is responsible for all the diversity of life we know of, from the first organisms like bacteria and green algae, to the millions of different species that inhabit our planet. From strange looking insects and breathtakingly beautiful plants and sea creatures, to elephants and ourselves. DNA is like a major revolution in the advance of complexity.

Whether it came from space or started on Earth, we know that life must first have occupied the sea or shallow lagoons because the land was still too inhospitable. Nor would the DNA revolution have taken off were it not for the formation of protective cell walls, which is another revolution in complexity. There is still controversy about how this happened but the skin of soapy bubbles caused naturally is one possibility. But once life had a foothold in the ocean it was destined to multiply. It is thought that huge masses of algae in the sea were largely responsible for making the free oxygen in our atmosphere. This increased as algae evolved into plants that then migrated from the sea onto the land, which allowed the first amphibians to emerge from the water.

Time Scale

Because we can now map changes in DNA it is possible to trace our ancestry right back to the first living organisms. Here are some of the milestones along the way. In fact, let's go right back to the beginning to give the whole picture:

Timeline

Quantum Vacuum		Zero Time
Big Bang	13.7 billion years	(13 700 000 000 years)
Earth forms	4.5 billion years	(4 500 000 000 years)
First fossilised algae	3.5 billion years	(3 500 000 000 years)
Primitive sea creatures	1.0 billion years	(1 000 000 000 years)
Vertebrates (fish with backbones)	470 million years	(470 000 000 years)
Amphibians (living in water & on land)	370 million years	(370 000 000 years)
Birds and 1st reptiles	320 million years	(320 000 000 years)
Age of the dinosaurs begins	250 million years	(250 000 000 years)
First mammals (animals with warm blood)	200 million years	(200 000 000 years)
Present continents take shape	100 million years	(100 000 000 years)
Dinosaurs become extinct	65 million years	(65 000 000 years)
Anthropoids (gorillas, chimps, apes)	35 million years	(35 000 000 years)
Hominids (ancestors of man)	5 million years	(5 000 000 years)
First use of fire	0.5 million years	(500 000 years)

Figure 2 Evolution of life

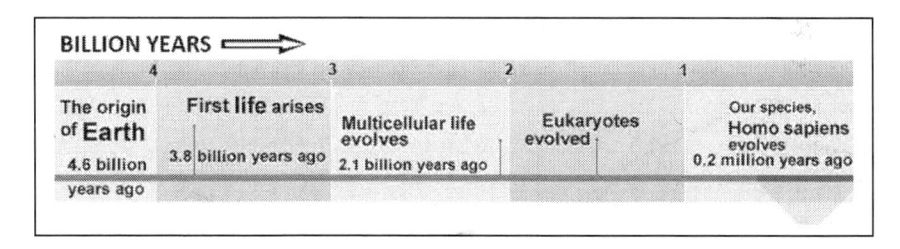

BILLION YEARS ⟹

4	3	2	1	
The origin of **Earth** 4.6 billion years ago	**First life arises** 3.8 billion years ago	Multicellular life evolves 2.1 billion years ago	Eukaryotes evolved	Our species, Homo sapiens evolves 0.2 million years ago

Note: Eukaryotes are organisms in which the cell nucleus is contained inside a membrane, i.e. all animal & plant life. We humans are eukaryotes.

Charles Darwin

The ancient Greeks had been puzzled when they found fossils, of what were pretty obviously some sort of shell-fish, high up in the mountains. That's because they had no idea how old the Earth was. By the nineteenth century many fossils were turning up in strange places, and a few people began to realise that life had been around for a huge amount of time. One of them was Charles Darwin's grandfather Erasmus.

This gave Charles a more realistic idea about the age of the Earth, unlike most people at the time who believed, on the evidence of the bible, that it was only six thousand years old. He went first to Edinburgh University to study medicine, and then changed to Cambridge where he developed an interest in geology which showed even more clearly that the Earth had to be of an immense age.

By remarkable coincidence in 1831 the captain of HMS Beagle, Robert FitzRoy, was looking for a 'gentleman's companion' to accompany him on a long voyage to the southern seas for the purpose of surveying the coastline for the government and navy. His first choice of companion backed out at the last minute, and Darwin, just twenty- two years old, was invited in his place. His job as the ship's botanist or 'naturalist' seems to have been almost an afterthought. The position was unpaid and his father thought it would be just a waste of time, but his mother came from the very wealthy family of Wedgwood chinaware, and it was Josiah Wedgwood who persuaded Darwin's father to let him go.

The voyage took them all around the coast of South America, including the Falkland Islands. Then round Cape Horn and into the Pacific as far as Chile and the Galapagos Islands, before crossing the Pacific to New Zealand and Australia, then on to Mauritius off the east coast of Africa, and around the Cape of Good Hope, before returning to England. The whole voyage took five years and, because the main purpose was surveying the coastline, the pace was fairly leisurely. This gave Darwin plenty of time to go ashore and explore. Along the way he was able to observe a huge variety of different species of plants and animals, including coral reefs, and even at one point an earthquake. High up in the Andes mountains he found geological evidence that the area had once been at sea level.

Figure 3 Charles Darwin.

Because they visited many islands, Darwin began to notice slight changes in the anatomy of the same species of animals on different islands. He noticed this particularly with finches that had different shaped beaks on different islands. And what was more, the shapes of the beaks were suited to the type of food which the island provided. He realised that it was the environment of each island that had somehow forced a change in the shape of their beaks over many generations.

It looked as though the finches had managed to adapt to their surroundings, or possibly, the ones that had not been able to adapt had died off, leaving the ones who could, to take over the island. Once the penny had dropped, so to speak, he began noticing how everywhere the design of animals had been brought about by a struggle for survival in competition for food resources in any particular surroundings. He called it 'Natural Selection', which was later dubbed 'the survival of the fittest' by critics.

Because of all the specimens he had collected and sent back during the voyage he became a well known naturalist on his return to Britain. For years he kept on doing research gathering more and more evidence from all over the world. He also bred his own pigeons to discover how selective breeding could change animals. Despite all this overwhelming evidence he didn't publish his theory for twenty years. Then out of the blue came a letter from a younger man named Alfred Russell Wallace who was working in Asian jungles. He had come up with the same theory as Darwin. Darwin then published his own work together with Wallace's. Darwin's book, "On the Origin of Species by Natural

Selection" was first published in 1859. It is widely considered one of the most important publications in the history of science. It gave a 'natural' as opposed to a religious explanation, as to how all life on Earth had come about.

You may wonder what became of Captain FitzRoy. He went on to become a pioneer in the science of meteorology. Using the best techniques then available he made the first weather predictions. He should have been hailed as a hero but, instead, whenever he got it wrong the politicians and newspapers made fun of him and mocked him to death – literally. He got so depressed that one Sunday morning, in despair, he cut his own throat!

Basically, all living things can only survive if they can obtain food. When there is plenty of food odd changes in individuals don't matter, but when food becomes scarce, those individuals who have gained a slight advantage in getting at the food (through chance genetic mutations), will survive and pass the changes on to their offspring. They will then multiply while the others will die off. In this way the changes for the better will multiply over time and give rise to different species. Currently there are thought to be a about two million species alive and kicking, though this doesn't include the likes of bacteria and microbial life of which there are a lot more, but the evidence shows that there have been at least two *billion* species since life began. Extinction is the norm, not the exception, and thank goodness, if it weren't we wouldn't have arrived.

Not Just Once

Nor has evolution only worked once. You might think it was a chance event, but it has happened time and time again on our planet. Each time evolving different kinds of animals and plants. There have been several mass extinctions which killed off nearly all of life on the planet. The worst was 250 million years ago, the most recent was 65 million years ago when the dinosaurs died off. But because of the mechanism of natural selection, life just seems to pick itself up again and create new creatures each time that are better adapted to the new surroundings, like the mammals taking over from the dinosaurs.

Darwin didn't know what created the slight changes in individuals, but now we know that it occurs because of random changes in our genes - odd changes in the DNA sequence that take place purely by blind chance. The great majority of changes are harmful, and those animals never grow old enough to reproduce before they die. But every now and then a positive change happens quite by chance, which gives an advantage to that individual, and he or she then passes it on to their offspring who are made more successful in surviving and reproducing, and so on. The H5N1 bird flu virus is a good example. Governments worldwide would not be so concerned if evolution wasn't deadly real. What they fear is a chance mutation happening to the virus so that it can spread between humans. At the moment it can only spread between birds (or to people handling sick birds), but as more and more birds are infected the virus is in seventh heaven and multiplies by the billion. Just a slightly nasty change in one of them, and you could have upwards of twenty million people dead. It happened in 1918.

More Evidence

One of the earliest examples of evolution in action for the Victorians was given by a species of moth in the Midlands of England. Before the Industrial Revolution it was well known as almost pure white. But with the pollution caused by the new factories that were built, it gradually became a much darker and dirtier colour – not because of the atmosphere, but because the ones with the darker colouring were less easily spotted by the birds, so they thrived at the expense of the whiter ones who were more easily seen and gobbled up.

Since then many more examples have come to light. The fastest ever example of evolutionary change was measured by a team from Harvard University. They studied a type of lizard that lives on six isolated islands in the Bahamas. What they did was to introduce a foreign predatory lizard. At first all the longer legged victims survived better because they could run faster. But within as little as six months the victims learned to spend more time up in bushes where the predators couldn't get to them. The ability to climb favoured the short legged victims so their offspring then began to dominate, leaving the long legged ones to be eaten.

One of the more convincing pieces of evidence that we are ourselves the result of evolution comes from something called 'atavism' which is when an individual's genes throw up a feature from our long distant past. There have been more than a hundred scientific reports of babies born with tails. Not as long as a monkey's of course, but quite long with several added vertebra and with the ligaments and muscle to go with it.
All of us as four week old embryos go through a stage of development when we have gills like a fish, revealing that in the distant past we evolved from creatures living in an aquatic environment. Later they turn into lungs. Babies born with webbed hands and feet are quite common, and one out of twenty humans has more than two nipples. As many as one in thirteen have flexible feet similar to a primate. Then there's the 'werewolf syndrome' where some people are covered in thick black hair like a chimp or gorilla. There are also those who are born with much bigger canine teeth than usual. This led Darwin to make the following rather caustic comment: "He who rejects with scorn the belief that the shape of his canines, and their occasional great development in other men, are due to our early forefathers having been provided with these formidable weapons, will probably reveal, by sneering, the line of his own descent."[2]

Evolution is by no means perfect. If it were we would surely have eyes in the back of our heads. We can see nothing behind us. It shocks me that the opponents of evolution are still arguing about the emergence of our camera like eye, when there is ample evidence that it evolved independently in many different species and at different times. But it's still one of the main starting points of the argument for Intelligent Design. What about the animals that have far better eyesight than humans, like some birds that can see four primary colours? Or bats with their highly developed radar? And what about the many different ways in which eyes have evolved over time? An earthworm has an eye, but it is just a bunch of light sensitive cells. Look at insects such as flies, with their mathematically shaped multiple eyes like the seeds of a sunflower, or the robotic stare of a spider's eight eyes.

And what about the selective breeding man has carried out on his own best friend, the dog? All dogs from the miniature Yorkie to the Great Dane have a common ancestor in the European wolf. In selecting for particular features that we rather like, instead of for survival, we have done exactly what evolution does naturally over a much longer period of time. The mechanism of both selective and natural evolution is staring us smack in the face. How can otherwise rational human beings reject evolution? It shows just how powerful the need is for people to have a belief in something.

If God put sea fossils high up in the mountains by magic to make us think the earth was older than it is, wouldn't that be pretty devious?

But now we have the final proof from DNA that we share a common ancestry with all life on Earth. You may find it amusing that 46% of our genes are exactly the same as those in yeast from which we make bread! And 50% of a humble banana's genes are identical to humans, and a little mouse has 85% of the same genes as us. Chimps have 98.7% identical genes to humans. Not believing in evolution is like asserting that the Earth is flat. Teaching children that evolution is only a theory is pure ignorance. To echo Steven Weinberg – for good people to perpetrate wickedness – that takes religion.

I cannot help thinking that the ultimate arrogance of our species is to assume that God intervened to make us somehow different and apart from the rest of Nature. As we'll be seeing on many occasions, Nature is far cleverer than the human mind can possibly imagine. What is even more unforgivable is that such arrogance makes us much less important than we *really* are.

Scientific knowledge has shown us quite unmistakably that we are the furthest development of matter in the universe, at least in this part of the galaxy. There can't possibly be a greater responsibility laid upon us as a species. To invoke a God who intervened in evolution, is to escape this huge responsibility. There is no longer the space for such arrogance. It is surely time now to buckle on our armour to confront this fearful destiny.

References

1. Gowers, T. 18 November 2006. What is Life. *New Scientist.* No2578. p 46-51.
2. Anonymous.13 January 2007.The Ancestor Within. *New Scientist.* No2586.p 28-33.

Chapter Three
Early Revolutions

Missing Links

So how did our species get here? The latest DNA evidence suggests that our line of descent branched off from our closest relatives the chimps between five and seven million years ago. Most scientists agree that around five million years ago an ape-like creature was forced out of the forests of East Africa, which were shrinking because of climate change, and onto the open grass lands. The landscape looked pretty much like it does today. They probably browsed and scavenged in the open and returned to the trees at night, much like monkeys and baboons do now.

The oldest fossil remains of an ape that walked upright on its hind legs, but still had tree climbing hands, dates back four and a half million years. It was discovered fairly recently in 1994. The next step towards modern humans arrived on the scene a million years later in the shape of what's known as the 'Lucy' skeleton found by Mary Leakey. Since Lucy, other similar examples have been found and given the name 'Australopithecus Afarensis'. Probably not a good subject to embark upon after a few glasses of wine! Its pelvis, leg and feet bones confirm that it walked upright. It also had a larger brain and its teeth and jaws were more modern.

Long before these finds, way back in 1924, Raymond Dart from the University of Witwatersrand had discovered Australopithecus Africanus, at a place called Taung in South Africa. This specimen was between two and three million years old, had a larger brain than Lucy, and also walked upright. It caused a lot of controversy at the time. Up until then, nearly all prehistoric remains had been found in Europe and the Far East. People didn't want to believe that there could possibly have been another species that walked upright, so long ago in backward darkest Africa. The British Empire was still largely intact at the time. The argument was finally settled rather surprisingly by a French Jesuit priest, Teilhard de Chardin, who was a world expert. He visited the site and confirmed Dart's findings.

Even after this there was still a reluctance to accept that humans first evolved in Africa. Robert Ardrey wrote a book about it called 'African Genesis' which became one of the first paperback popular science books in the late 1960's. Because of all the later finds in East and North Africa it is now accepted that humans first came from Africa. However, just recently the oldest primate ever discovered, called Archicebus Achilles, was found in Eastern China and dates back 55 million years. It is a lot older than the oldest primate from Africa, so the quest for our origins has taken an unexpected turn.

In terms of evolution, Dart's find was followed by another Australopithecus called Robustus who had an even larger brain of 530cc.compared with ours which averages 1 350cc. It also had a larger body. You will have gathered that all these 'Australopithecus' specimens had nothing whatever to do with Australia despite the sound of the name. Robustus lived between two, and one and a half million years ago. I should say that I am only covering the major milestones. There have been several more actual finds, they occur fairly frequently and are slotted into the main framework and given another Latin name.

First Tools, Speech and Fire

Alongside Australopithecus between two and a half million, and one and a half million years ago, lived Homo Habilis, first found by Richard Leakey (son of Mary Leakey) at Olduvai Gorge in Tanzania. Other remains have since been found in Ethiopia. This was the first proper toolmaker. Tools were made in the usual way by striking two stones together to get a rough edge which could be used for cutting and as a weapon in hunting, and no doubt in disposing of unwanted visitors. But these first stone tools were very crude in comparison to what came later. To give credit where it is due, these tools represent a big evolutionary advance, and could perhaps be considered the first 'technological' revolution. It represents another milestone in the advance of complexity.

But the next character was even more impressive. Known as Homo Erectus, they were contemporaries of Homo Habilis and probably lived along side each other. The earliest finds start two million years ago and end only a recent half a million years ago. Surprisingly, they lasted much longer than modern man has so far. Although they still had thick brow ridges, they did have a much bigger brain at 1 200cc. Because of the way their tongues were fixed in the skull, it is thought that they may well have had speech.

This must surely be one of the most important advances in complexity that has happened up to this point. Once you can say or 'articulate' the thoughts in your head to another consciousness, then you can communicate more complex ideas and the effect is cumulative. Much has been written about the influence of language on brain development. Their vocabulary was probably pretty basic but it would have made them far more effective as hunters, particularly where careful co-ordination between groups was necessary when trying to hunt animals. The evolution of speech is another major revolutionary advance.

Their most important contribution in a technological sense was the domestication of fire. This is another big revolution. It brought them heat and light at night, and allowed them to travel to colder, less hospitable climates. It must also have given them increased protection from wild animals, and improved their diet by being able to cook food.

Equipped with these innovations they were able to migrate out of Africa and over most of the world with the exception of the Americas and Australia. There have been many finds, starting with 'Java Man'(*Pithecanthropus*) in 1893. Others were 'Heidelberg Man'(*Homo heidelbergensis*), 'Peking Man'(*Sinanthropus*) in China, and 'Turkana Boy'. They made much more sophisticated tools, including wooden bowls and hand axes. There is also some evidence that they built shelters and may have even had a musical instrument in the shape of a bone flute.

They died off only half a million years ago. It would be a mistake to think that they were human, but they were not far off.

Modern Humans

The oldest remains of modern humans so far found, date back a hundred and ninety thousand years and were found near the Omo River in Ethiopia. Quite amazingly, by analysing a certain section of DNA known as the mitochondria, it is now possible to trace all humans alive today to a single female ancestor who lived a hundred and fifty thousand years ago. She is referred to as Mitochondrial Eve. Such is the power of DNA analysis.

I haven't forgotten the Neanderthals. We parted from the same branch as them about seven hundred thousand years ago. They are considered the same species as us and are officially called Homo Sapiens Neanderthalis, whilst we have given ourselves the much better sounding title of Homo Sapiens Sapiens. My schoolboy Latin tells me that sapiens means 'wise'. I don't know why they repeat it twice. It seems an odd sort of double assertion. And certainly fairly dubious, considering the history of the twentieth century.

Neanderthals have had a bad press over the years, but the truth is that they had a bigger brain than ours, at 1 500cc compared to 1 350cc. However, they were somewhat vertically challenged coming in around five foot five inches and having a much heavier and slightly stooping frame. They had similar tongue attachments so could speak, but their tools and implements were not as refined or as specialised as Sapiens Sapiens.

Their remains have been found in parts of North Africa and the Middle East, and they are known to have arrived in Europe about 120 000 years ago. They buried their dead, which is an innovation that seems to add something to the quality of their consciousness. The burials appear to have involved a certain amount of ceremony and ritual which illustrates a degree of 'concern', even perhaps thoughts about an afterlife. In one case a youth was buried surrounded by a ring of horns, and in Northern Iraq another was laid on a bed of masses of wild flowers and grasses. Another grave revealed a man who had lost an arm many years before he died, suggesting that they had a certain amount of compassion for the less able.

Cro-Magnon

About twenty thousand years later (100 000 years ago), Sapiens Sapiens migrated out of Africa and into the Middle East where Israel, Lebanon and Turkey are today. Roughly 50 000 years ago, they moved north into Eastern and then Western Europe, and joined the Neanderthals. They also spread east into Eurasia, and south and eastwards through India, arriving in China around 80 000 years ago. They reached Australia about 35 000 years ago, and crossed the Bering Strait into North America from Russia perhaps 15 000 years ago, although the timing is still disputed.

In Europe they lived alongside the Neanderthals for some 20 000 years, an enormous amount of time when you think that the Romans were building their empire a mere 2 000 years ago. There is now conclusive DNA evidence that we interbred with Neanderthals. Europeans today share a small percentage of their genes with their ancient cousins, whereas people of sub-Sahara Africa do not. The most surprising recent development has been the discovery of another line of humans never known before. By sequencing the genes in bone fragments found in a cave in Siberia they have established that these people, known as the Denisovans, shared parts of Russia with both Neanderthals and modern humans about 40 000 years ago. Remarkably, about 5% of Denisovan genes still survive in the people of Papua New Guinea and the Aborigines of Australia.

Yet another human line, known as Homo Floresiensis, was discovered on the Indonesian island of Flores. They were very small in stature so are more often referred to as the Hobbit. The remains date back to around 17 000 years ago. There is still controversy about whether they are a separate line of humans. Even more recently another line of humans may have been found in China. They are known as the Red Deer Cave people, but a way has not yet been found to sequence their DNA. So what palaeontology is showing us is that we are the last surviving group of humans. Our lineage slit from chimps, gorillas and orang-utans more than five million years ago, so it is unlikely that evolution will contrive another intelligent species any time soon.

In Europe Sapiens Sapiens is called Cro-Magnon, after one of the first sites where they were found. They had smaller faces, a lighter skull and straighter limbs than Neanderthals. In fact they looked just like us, except that evolution has worked on us in that short time to give us smaller molar teeth and a skeleton that is about 15% lighter. Like our predecessors we were hunter gatherers living in small groups as nomads, consisting of a few families each, having contact with other nearby groups like Amazon tribes do today, or the Bushmen of the Kalahari in Botswana. This will become a more important consideration later, because the size of human groups may be a key to what our own future holds.

Cro-Magnon had a much more sophisticated consciousness. We can infer this from several things. They buried their dead with necklaces of shells and beads and with their favoured weapons. Their tools were highly specialised, and included finely made bone harpoons for catching fish, and bone needles to sew animal skins together for various uses such as clothing and tents, much like the Inuit today. Their spearheads were made of sharpened antler and were balanced for throwing. In the later periods they used bows and arrows.

Perhaps their most celebrated advance was the art they produced. I like to think that Cro-Magnon art is important because it is one of the earliest examples of non-verbal information transfer. First from the live animals in the wild, then to a human consciousness and finally to a cave wall. It adds another layer of complexity to the planet. It tells us that they had a consciousness, an 'awareness' if you like, that was the same as ours. The famous cave paintings at Lascaux in France and Altimira in Spain are quite breathtaking in their skill and precision. Their depictions of animals long since extinct are a direct window into their world. So beautiful they make a lot of subsequent art look a bit silly.

But they also left behind many engravings and sculptures. These included small clay figures of large breasted females, which they obviously thought was important. Strange how, across the ages, not much has changed considering the current popularity for silicon implants. But they also went in for bone and ivory carvings. They decorated their tools and weapons with animal designs and intricate patterns. This was a much richer and more complex culture than anything that had evolved before. Modern Humans had arrived.

Chapter Four
The First Civilisation

The Farming Revolution

The last ice age ended ten thousand years ago, and with minor ups and downs in climate the world has fortunately remained comfortable enough for us to survive as a species. We are extremely lucky for this very brief interval which is still continuing. There have been perhaps twenty ice ages in the last two million years caused, it is thought, by the Earth tipping on its axis first towards and then away from the sun in its otherwise steady orbit. That's how delicately we are balanced and how precarious is our existence. And two million years, as we have seen, is actually no more than the space of one breath in terms of the major changes that have taken place in the past.

As the glaciers retreated, one of the greatest revolutions of all in our brief history began to take place. It didn't happen overnight but it was a major change that provided the foundation on which civilisation could take place.

It's usually called the Agrarian Revolution, or the invention of farming. This can perhaps sound a bit boring to present day city dwellers but it is actually an event of enormous importance. There are still a few groups of nomads and hunter-gatherers around who have not yet made the change to farming. Like the San people of the Kalahari, the remote tribes of Amazonian Indians, the Bedouins, and the people of the Mongolian steppe. The first farmers are known as the Natufians who date back some 13 000 years.

No one can be sure exactly why or how it happened, but the domestication of crops and animals happened in several parts of the world that all had the same conditions to make it possible. These were large river valleys where there was a constant supply of water and soft fertile soil that could be worked quite easily with a wooden plough. One of the first areas where it happened was the land along the banks of the Tigris and Euphrates rivers in Mesopotamia which is where Iraq is today. Another area was along the river Nile in what was to become Egypt. The land between the Mediterranean Sea and the Jordan River was another. In India it occurred along the valley of the Indus River, and in China the Hwang-Ho (Yellow) River.

What started it was probably over-population. As hunter gatherers, a group of four or five families living together would have required at least a few thousand acres of land to survive by chasing game, but if you could grow crops and herd animals you only need a few acres. We know that wild barley grew in southern Turkey, and wild wheat in the Jordan valley. Chickpeas which are highly nutritional grew in Mesopotamia at this time, and some experts have suggested that this is the reason why civilisation first began there. The area was more lush just after the ice age than it is to day, and the whole area from Egypt to Syria and Mesopotamia is sometimes referred to as the 'Fertile Crescent'. In India and China there was wild millet and rice.

It is easy to imagine, probably the women, gathering these wild cereals while the men went out to hunt for meat, as happens in modern hunter-gatherer societies. It's inevitable that some of the cereal they gathered was spilled on the ground and left behind as they followed the game. When they returned, perhaps the following year to the same camp site, they noticed that what they had left behind had grown up and was ready to hand, so they didn't have to go out and gather it.

Bladed tools for harvesting cereals have been found in southwest Iran from ten thousand years ago. They were made of obsidian, a type of volcanic rock that can be flaked to make long blade edges. The really surprising thing is that there is no obsidian in Iran. The nearest source is southern Turkey. This suggests that they were already trading goods over several hundred miles.

In a similar way, from hunting game over large areas, and having to move camp every few days, there were obvious advantages to herding animals instead of chasing them. The first traces of keeping sheep come from Northern Iraq at this same period. Later evidence shows that they had goats and pigs. Once they could grow their own cereals and herd animals there was no longer any need to keep moving. The result was that settlements grew up and became permanent.

The First Town on Earth

The earliest example of a village is at Jericho, not far from Jerusalem. It's still a thriving place today. There has been a settlement there for eleven thousand years, probably because it has a spring that has never dried up. By ten thousand years ago there were several acres of mud brick houses. Later it supported about three thousand people. It also had a stone tower and defences so it could legitimately be called the first town on Earth. Even earlier (13 000 years ago), the remains of the first temple have been found at a place called Göbekli Tepe in Southern Turkey, and in Jordan at a place called Wadi Faynan there is what appears to have been an amphitheatre which has been dated to 13 600 years ago. So everywhere along the river valleys, villages and small towns began to pop up. They had even more refined stone and bone tools, and made decorated pottery out of clay by hand. With keeping sheep and goats, there was an immediate supply of skins and wool for clothing as well as dairy produce.

What's so significant about the invention of farming then? The most obvious is that it provided an economic surplus. It meant that not everybody had to be involved in the production of food. But if we follow the theme I am suggesting, then of equal importance was the fact that more people could live in a smaller area and in larger groups, which meant bringing together more human minds that could interact with each other and create new ideas. If you think of us as just 'matter' getting slowly more complex from the Big Bang onwards, then farming has added an important layer.

Put another way, it means that once there was surplus food, it gave people time to do other things like just sit about, share some liquid stimulant together perhaps, and swap ideas The earliest evidence of wine being made dates back 7 100 years.

Metals

Between eight and nine thousand years ago the first evidence of smelting copper appears in Turkey. Copper is too soft on its own to be of much use other than as ornament or for domestic use, but by six thousand years ago they had learned how to blend copper and tin together to make bronze which was harder and more useful. So over a period of time the Stone Age passed away.

Making bronze is quite a stride in technological development, and shows how important it was to first produce surplus food before advances like this could be made. With the arrival of the Bronze Age the stage is set, and all the ingredients are in place, for the emergence of the first civilisation.

Sumerian Civilisation

By seven thousand years ago there were hundreds of farming villages dotted all around the Middle East, all providing surpluses and trading with each other. After bronze was invented the areas where the copper and tin were found also became settlements in their own right. Some of these villages, as we have seen, grew into towns, probably with their own cults and religious practices.

Around this time pottery becomes duller and less interesting but more plain and uniform. Someone had invented the potter's wheel, and pottery was being mass produced. Another technology had been developed, providing objects for trade and more links for communication. As time went by a few of the towns grew into city states, each with its outlying towns and villages. They would have needed some form of administration which means that different social classes probably arose at this time. However, this is not yet what you could call a civilisation.

The first proper civilisation began in Mesopotamia, now called Al Jazeera. The current TV station reflects this. It is the area around the Tigris and Euphrates rivers where Iraq is to day, and is known as the Sumerian civilisation. It lasted thirteen hundred years, which is a substantial achievement. If we go back the same amount of time from today, we end up in the European Dark Ages after the fall of the Roman Empire. So it lasted quite a long time.

It started with city states which then came under the government of a single ruler. Three of the main cities were Ur and Uruk (possibly where the modern day name of Iraq comes from) in the south and Adab in the north. Nippur, just south of where Baghdad is now, was a big religious centre. Other cities were Kish and Lagash. Towards the final period some of the cities grew to as many as sixty thousand people.

So what makes a civilisation different from a city state? The most obvious, as I have said, is that it consists of several cities all under one ruler or ruling family. All the people share a language and a religion and have a common distinct culture. Another prominent feature is big buildings in the form of palaces and temples. The oldest surviving monumental building in the world is the Ziggurat at Ur which was over a hundred feet (thirty metres) tall. Most later civilisations follow the Sumerian model. What they represent is a bigger area with a greater number of people living and working together in an orderly fashion under a common set of rules and regulations. Notice that with a common language and religion, there are now far more human minds communicating with each other than ever before.

Invention of Writing

With this set up another big revolution takes place and with it another level of complexity is added. The invention of writing was like the first internet. It allowed the transmission of ideas and information over great distances and over long periods of time. Without it the organisation and administration of a group of city states would have been much more difficult.

THE FIRST CIVILISATION

Without a doubt, one the greatest achievements of the first civilisation was the invention of writing which happened around 6 000 years ago. It started in a crude way of course, with simple pictures representing things, but gradually gets more and more complicated until single signs come to represent whole words. The Sumerians used clay tablets to write on. They did this when the clay was soft by chopping off the end of a reed to give it a sharp edge so that they could make marks in the clay. The clay was then baked hard. A great number of these tablets have been left behind so that we can get a very good idea of how they lived. Their writing is known as cuneiform.

These tablets reveal that there was a single ruler or king, surrounded by a royal family and a nobility, a religious caste of priests, and a professional military. Under these came the craftsmen and traders, and then a mass of ordinary farmers and labourers, and a caste of slaves.

It seems that the first use of writing was economic. It was used to record things, like lists of goods and receipts, and later to record events like wars and famous victories. The very first great story, in fact the first literature in the world, is known as the Epic of Gilgamesh, which is about an actual king. It tells of a great flood, just like the story of Noah, but several centuries earlier. It tells how the gods send a great flood of water to destroy everyone because of their wickedness, except for a single family who are saved by building an ark would you believe? When the flood subsides they multiply and become the chosen people. The similarity to the biblical story of Noah's ark is striking. It shows how much the Sumerian civilisation influenced much later traditions, right up to many a Sunday school today. Abraham, the founder of the Jewish religion, is said to have come from either Uruk or Ur around 2 100 BC. Scholars have also proposed that the Old Testament story of the Tower of Babel could be a reference to the Ziggurat at Ur.

Gilgamesh is the world's first heroic character who commits great deeds of bravery and daring. The roots of many other later stories also appear for the first time in this epic. It includes the story of a wild man called Enkidu who was pure and innocent because he was raised in the forest by animals, much as Romulus and Remus (the mythical founders of

Rome) were, much later. When he grows up he helps the animals by destroying the traps laid for them by man. The trappers get mad about this, and go to see king Gilgamesh. He suggests that they persuade the beautiful and enticing Shamat to corrupt him, and lure him way from his innocence in the forests. This bears more than a passing resemblance to our later story of Adam and Eve. One big difference being that Shamat is a temple harlot, which also confirms the old adage about the oldest profession on Earth.

She meets up with Enkidu and they have seven days and seven nights of lustful passion after which Enkidu has lost his innocence and agrees to leave the forest and become civilised. He goes looking for king Gilgamesh and they have a terrific fight, after which they become the best of friends. They go on many adventures together. One of them is to the land of Lebanon where they fight and kill a monster with the wonderful name of Humbaba, (shades of Jaba the Hut from Star Wars). Later the beautiful goddess Ishtar falls in love with Gilgamesh but he's not having any of it, so she kills his pal Enkidu in revenge. A ripping story, and that's just part of it.

Figure 4 Goddess Ishtar.

Another great bit of writing comes from a high priestess of the moon god Inanna. Her name was Enheduanna and she was the daughter of Sargon, king of Akkad, who we will hear more about in chapter five. She wrote forty-two hymns to the moon god. Apart from being the first author actually known by name, her work is the first record of a human giving an account of the inner life of their consciousness. Another layer of complexity perhaps?

Religion and Laws

The Sumerians also had three main gods. Anu, who was god the father, Entil, lord of the air, and Enki, god of wisdom. Not entirely unlike the Christian Trinity. The Sumerians were also responsible for the first recorded legal system. It laid down laws and codes of behaviour for all citizens. For example, marriage was only allowed with the consent of the bride's family and had to be monogamous, which is fairly amazing considering there are still many societies that practise polygamy. Marriages were recorded and sealed with a contract. It even allowed for divorce and protected the rights of women. Just like other generations ever since, they also worried about the youth of the day getting out of hand. A clay tablet found at Ur says, "If the unheard of actions of today's youth are allowed to continue, then we are doomed." What's new?

The Wheel

One of their most useful innovations was the invention of the wheel for transport. We tend to take this round shape for granted without thinking much about what an advance it represents, but it's another revolution. It seems from the evidence, that the potter's wheel actually came quite a long time before the idea was transferred to wagons and chariots. Perhaps a couple of inventive builders, who had a pile of mud bricks to move, were sitting discussing their problem in a local tavern one day, when in walked a potter and they bought him a drink? We'll never know.

What we do know for sure is that great civilisations like the Maya and Inca of south and central America never came up with the idea, nor did the north American Indians or the tribes of sub-Saharan Africa.

Commerce and Art

The Sumerians were also great at commerce. We know that they traded as far north as Syria and Turkey, and as far south as Iran and the Persian Gulf. There is even evidence that they traded as far away as the Indus River in India. Their art was also quite advanced. They were the first to create plausible representations of humans in engravings and sculptures, so that we know what they looked like and how they dressed. For instance, it shows us that most of the men went in for what we now call 'skin heads' but with the addition of a big beard. It could even catch on today. From the figures they have left us, which show very prominent eyes, it seems that both sexes rather overdid the eye make-up by today's standards. Their jewellery was intricate and elaborate, with lapis lazuli being considered particularly chic. It was obviously thought to have quite a value because it was traded from beyond their borders. They also worked in gold and silver.

Numbers

From the point of view of layers of increasing complexity, their greatest discovery has to be the use of a number system. It most likely grew from the simple need to keep accounts and record things, like the number of pots our friend had made for the temple that day, or the number of sheep a farmer sold at the market. By using a series of indents in a soft tablet of clay to represent a number, instead of using pebbles or sticks to count, they discovered arithmetic, where you can 'times things' in a very short space, instead of making several new piles of stones. This grew from recording the number of animals a farmer had, to the more sophisticated measurements required to erect monumental buildings. It should also be said that they were great engineers. Because the river deltas frequently flooded, they built large embankments and dug canals and irrigation channels to water the surrounding land.

Perhaps the closest we can come in our every day lives to feel the presence of the people of Sumer, is in their discovery of numbers. They used a system where the number sixty was the main base. Just as our main base is a unit of ten, theirs was sixty. From them we inherited our sixty seconds in every minute and sixty minutes in every hour, and the twenty four hours for each day and twelve months in the year. And the way any circle is still conveniently divided up into 360 degrees. Each time we glance at a clock face or our watches, we share something with those people who lived five and a half thousand years ago.

Their temple priests were the first people we know of to actually record astronomical events, and they were also able to observe the movements of the planets against the fixed stars, and came up with the first recorded calendar. Using this, they were able to accurately predict the seasons and the lunar month. The trouble was that they believed that the stars influenced human behaviour, so they were astrologists rather than astronomers.

The history of Sumer divides roughly into four main periods.

The Pre-dynastic or Archaic 2900 – 2750 BC
The 1st Dynasty at Ur 2750 - 2340 BC
Sargon I Akkadian period 2340 – 2100 BC
The 3rd Dynasty 2100 – 1800 BC

References

1. Reno, J., Springen, K., Meadows, S., Underwood, A., and Scelfo, J. March 5, 2007. Girls Gone Bad. *Newsweek.* Vol.CXLIX, No.10. p 50-53.

Chapter Five
Civilizations of the Middle East

Hammurabi

Sumerian civilization comes to an end when the Elamites destroy Akkad around 2000 BC. A period of some confusion follows until Hammurabi, the sixth king of Babylon, comes to the throne in 1764 BC. He manages to subdue the whole of Mesopotamia and establish the first Babylonian Empire. He is important because from him we have the oldest surviving written code of laws.

Because it is the origin of written law, copies of Hammurabi's seal are often found in government buildings around the world, including the Supreme Court of the United States and in the House of Representatives. His 282 laws were carved into a slab of basalt rock about eight feet (2.4 metres) tall with a picture of Hammurabi at the top receiving the laws from a god. This implied that the laws didn't come from the king but from a higher authority. There may have been one of them in every town, probably outside the temple for all to read so that no one could plead ignorance of the law. The only surviving copy is in the Louvre in Paris.

It listed crimes and their punishments which were pretty harsh by today's standards. There were also ways for people to settle disputes and guidelines for everyone's behaviour. I want to look a bit more closely at their family laws because I find them fascinating. But if you are not particularly interested, just push on to the next section.

Family Laws

Marriage required a contract between man and wife that was legally binding. The ceremony even included the 'joining of hands' and required the man to make a pledge in public to look after the wife. Marriages were arranged between families, and the groom's father had to provide the 'bride price' which the groom gave to the bride's father who then, surprisingly, had to give it to the bride. If, after he accepted the presents, he didn't give up his daughter, then he had to repay the presents to double the value.

If the groom changed his mind at the last minute, he had to forfeit all the presents and give them back. The dowry that came in addition to the presents, didn't go to the groom and his family, it belonged to the wife for life and could be passed on to her children. Once married, the couple were considered a legal unit and the man was responsible for the wife's debts, even those she had before the marriage, as well as his own. Both were responsible for all debts after the marriage. Despite being married the bride always belonged to her father's house.

Divorce was optional for the man but it didn't come lightly. He had to give back the dowry, and any children automatically became the custody of the wife. The ex then had to assign the income from a field or garden to her, as well as goods for the maintenance of the children until they were grown up. But the wife was free to marry again. If there were no children the husband had to return the dowry and bride price. The bride price was added in compensation for her rejection.

Court actions were obviously a serious business. The wife could bring an action against the husband for cruelty and neglect and, if proved, she could obtain a judicial separation taking with her the dowry. No other punishment fell on the man. But if she didn't prove her case, and was shown to be a bad wife, she was drowned! Another curious arrangement was that if she was left without maintenance for a long time by the husband doing military service, or being away on business, she could cohabit with another man until his return. The children of the second union then stayed with the father. If he deliberately deserted his wife the marriage dissolved, and should he subsequently return he had no claim on her property.

When the husband died the wife took over all his property and was responsible for bringing up the children. She could only remarry with the consent of the court, and the judge had to make an inventory of all her property before handing it over to her and the new husband to keep in trust for the children. Monogamy was the rule but, if the union proved childless, the wife could give her husband a slave maid to bear him children who were then recognised as the wife's. She remained mistress of the maid and could downgrade her to a slave again for insolence, but could not sell her if she had borne children to her husband. It is acknowledged that one of the patriarchs in the bible had a child by a similar Jewish convention.

These laws give an intriguing insight into how the society worked, and a level of sophistication not matched by some modern cultures. They also set a precedent for carving laws into stone that was followed elsewhere, and survives perhaps in English with the expression 'written in stone', to denote something definite. The Hittites, who we come to shortly, had laws written in stone from 1300 BC. A more familiar example comes to us from the law of the Hebrew patriarch Moses with the Ten Commandments which are thought to date from around 1200 BC.

The Egyptians

About four hundred years after the beginning of the Sumerian civilisation, the northern and southern kingdoms of Egypt were united under a king called Menes. This took place in 3200 BC. It marks the beginning of Egyptian civilisation which was to last for three thousand years, well into the classical period of Greece and Rome. Around 2500 BC the Harapan civilisation begins in India, and in China civilisation starts with the Xia and Shang dynasties around 2000 BC.

The story of Egypt is closely tied up with the river Nile which overflowed its banks every year for about two months. Each year the river brought new sediment that made the earth very fertile, but they had to construct big reservoirs and irrigation canals to keep up the supply of water over the dry season. This required a strong central government that could command a mass of manpower. Perhaps this is one reason the Pharaoh's became god kings.

They were associated with Ra the Sun God and were thought to be divine. Hence the construction of the pyramids as monumental buildings for the dead. In sheer size not much has ever surpassed them except perhaps the Great Wall of China and the burial chambers of the first Chinese Emperor Qin Shi Huang (210 BC), whose huge necropolis containing the Terracotta Army has not yet been fully excavated.

Unless, that is, you count the Large Hadron Collider at the European Centre for Nuclear Research (CERN) near Geneva. It is the biggest machine ever built, and has a large underground tunnel twenty-seven kilometres in circumference crammed with over a thousand huge magnets that have to be super cooled, and detectors as tall as a four storey buildings. It is a machine built by an international partnership with the intention of furthering our knowledge of how the universe works.

In Egypt they started building the first pyramids at Giza in 2700 BC, and the two biggest, the one for Kufu (Cheops) and the one for Chephren, were constructed over sixty years between 2550 to 2490 BC. In addition to the pyramids there are also the towering columns of the great temples at Karnak and Luxor, as well as colossal statues like the Sphinx and the seventy foot (21 metres) tall image of Rameses II built in 1255 BC. Then there are also the many tombs in the Valley of the Kings. The Egyptians were also responsible for the invention of using papyrus reeds to make parchment to write on which was an advance on clay tablets that were bulky and heavy to carry around.

They were also wonderful artists. Think of the paintings of river scenes with boats and fishermen, or peasants harvesting crops. The images also show us a rich lifestyle that includes scenes of slaves pouring olive oil or perhaps wine from big jars at banquets complete with musicians and naked dancing girls.

Figure 5 Egyptian lifestyle

Their art represented fine portrait sculpture of actual individuals which is perhaps another layer of complexity in itself. Think also of the fabulous jewellery and ornaments, and the gold and silver ware. A good example is the treasure of the Boy King Tutankhamen who reigned from 1347 to 1339 BC. Some sources think that it was during the reign of Rameses III (1182 – 1151BC) that the biblical Exodus might have taken place when Moses led the Hebrews out of slavery by allegedly parting the Red Sea.

A hundred and thirty years before Tutankhamon, Thutmose III came to the throne. He ruled for fifty-four years and went on something like seventeen war campaigns, creating the biggest empire in Egyptian history. It stretched far to the south along the Nile, including Nubia, and to the North the Arabian peninsular, the whole of the Mediterranean coast where Israel and Lebanon are to day, and right up to where modern day Syria is. Unfortunately this caused a head-on collision with a people who were conquering southwards.

The Iron Age Begins

These folk were known as the Hittites. They appear quite often in the Old Testament. They came from beyond the Black and Caspian seas and settled to begin with in Turkey. It's here that the first evidence of iron weapons and implements crops up, but about the same time it also appears in India. Because iron is much harder and more durable it gave them a big advantage. Some scholars think that the Hittite kings kept it a secret to prevent anyone else from learning how it was made.

Extracting iron from iron ore is a complicated technology and represents a major revolution. You need a much hotter heat source than an ordinary fire. You can only get these temperatures by using charcoal, and you have to keep up the heat for several hours. Not only that, you have to reheat it, and reheat it again, to get rid of impurities and then add carbon to make the iron harder.

Before coming up against the Egyptians, the Hittites conquered the whole of Mesopotamia. Over the years their struggles with the Egyptians went on and on, and eventually they decided to make a peace treaty in 1280 BC. It's the first evidence of such an agreement between empires and it worked because the Egyptians later came to the aid of the Hittites. By the time of the treaty, the secret of how to smelt iron was out and its use had become widespread. The Bronze Age like the Stone Age before it had passed away.

Because the Egyptian and Hittite empires fought themselves to a standstill, so to speak, it left a power vacuum in the land between them along the Mediterranean coast and inland to the river Jordan. It's thought that around 1500 BC the Israelites, who were a nomadic tribe up to this point, wandered into the region. At the time the Egyptians were still in control of the area and many of the Hebrews may have gone down into Egypt. Whether this was by way of selling their brother Joseph into slavery there is no way of telling. But about two hundred years later they were established in the area west of the Jordan River and had given up the nomadic life, which could have been the end of the Exodus.

When they arrived in the promised land, they came up against the Canaanites who were already there and managed to subjugate them. They in turn were defeated by the Philistines. But then came the legendary fight between David and Goliath. The Philistines were sent fleeing and David became king of Juda. It was during the reign of his son Solomon that the Hebrew kingdom reached its height. He built the famous temple in Jerusalem and probably held sway over quite an area from the sea to the Jordan valley. The Old Testament tells us he had a reputation for being a good ruler and very wise. Remember the beautiful Queen of Sheba from the south who came bearing gifts to witness the wisdom of Solomon? It could well be true.

About the same time the Phoenicians or 'Sea people' from the coast of Turkey began to branch out. They were traders and colonisers who built the cities of Tyre, Sidon and Biblos on the Eastern Mediterranean coast. Over a period of time they came to control large stretches of the Mediterranean coast as far away as Spain, but their main centre became Carthage on the North coast of Africa near where Tunis is today. They lasted a long time. Right through to the Carthaginian wars between Hannibal and the Romans which started in 264 BC.

It was the Phoenicians who discovered how to make glass. They also adopted a combination of Egyptian and Babylonian writing and developed it so that it had an alphabet of just twenty-two consonants. When the Greeks had had a lot of contact with them through trade, they decided to take up their system of writing and so it was imported with many later changes into Western Civilization. It was because Egyptian papyrus was traded through the Phoenician port of Biblos that written works became known as biblia. So the word passed from the Phoenicians to the Greeks and then on into Latin and into 'bible' and 'bibliography' in English.

The next important figure is Sargon II of Assyria (772 – 705 BC). He conquered all of Mesopotamia and Palestine and occupied Egypt. His successor, Sennacherib, deported some 80 000 Hebrews to Mesopotamia. Incidentally, he is credited with starting the first postal service to maintain contact with all his provinces. But by 612 BC a new people, the Medes and Persians, join up with the Babylonians to defeat the Assyrians and capture their capital Nineveh which we hear about in the Bible.

After this, the Babylonians enjoy a second period of supremacy under their king Nebuchadnezzar (604 – 561 BC). During his reign Babylon reaches its greatest glory as a city. Yet again Jerusalem is attacked and the Jews are taken into exile. This is the period referred to in the Bible as the Babylonian Captivity. However, then comes the famous feast of Nebuchadnezzar when, as the Old Testament story goes, the king saw the writing on the wall, followed shortly after by the arrival of the Persians under Cyrus the Great. He and his successor go on to create the biggest empire thus far. It stretched from Egypt in the south through where Israel/Palestine and Lebanon are, to include all of Turkey and Mesopotamia right up to northern India.

The empire was efficiently run. It was divided up into twenty provinces, each governed by a Satrap whose power was limited by a secretary and a military commander who both reported directly to the king. Different peoples were allowed to follow their own religion and local customs so that life went on as usual. It was Cyrus who allowed the Hebrews to return to Jerusalem and rebuild their temple.

The Minoans

The Minoans, the first civilized people in Europe, lived on the island of Crete. No one is sure where they came from but it was probably somewhere in the eastern Mediterranean. Although they had their own distinctive language and writing, no one has yet been able to decipher it. I'm always surprised by this, but by 1900 BC they had constructed a sophisticated society with harbours and towns and huge palaces with hundreds of rooms. The palaces had separate areas for different crafts like pottery and weaving as well as metal workshops. They were more like a small town under one roof so to speak.

An unusual feature was that there were absolutely no fortifications or weapons found in the palaces. This suggests that they had such a complete command of the seas around their shores that they never brought weapons inland. The biggest palace was at Knossos, but there were others at Festos, Malia and Zakros.

In 1700 BC disaster struck in the shape of an earthquake and all the palaces were destroyed. But this was the height of the Minoan civilization and they rebuilt them on an even grander scale. These are the remains that can be seen on Crete today. Usually built around a courtyard surrounded by colonnades with staircases and hundreds of rooms leading off. The walls are elaborately decorated with frescoes of daily life depicting feasts with dancing and music, as well as fishing and harvesting. From the wall scenes and from sculptures it seems the predominant fashion for women was a form of 'topless dress' which left them bare breasted.

It is thought that they had some kind of 'bull worship' probably for reasons of fertility. Bulls' horns adorn the decorations and many sculptures and clay models of bulls have been found. Some of the frescoes also show naked boys and girls in a sort of gymnastic ritual where they grabbed the bull by its horns, leaping over its back.

This bull worship may have given rise to the later Greek legend of King Minos and the mysterious labyrinth harbouring the awesome Minotaur which was half-man and half-bull.

In 1450 BC disaster struck again but this time it wasn't an earthquake. The palaces and private villas were destroyed by fire but the towns left unharmed. This suggests that it might have been an uprising of the ordinary people against the ruling class, or possibly an invasion.

The Mycenaeans

The invaders may have been the Mycenaeans. They had arrived in mainland Greece earlier, around 2000 BC, and built up a number of city-states along the coast. The most important of these was called Mycenae, hence their name. They also founded other cities. The most famous was Troy on the coast of northern Turkey. These were the first Greek speaking people. Because of the mountainous coastline with its rugged inlets and natural harbours, it was easier for the city states to communicate by sea, which led them to become seafaring. As a result they met up with the Phoenicians and absorbed their system of writing. It is the Mycenaean's that the famous Greek poet Homer writes about in the 'Iliad' and 'Odyssey'. The famous Trojan War, fought over the abduction of the beautiful Helen, whose face is said to have launched a thousand ships, and the clever trick of the giant wooden horse, actually took place in 1200 BC. Although this was long before Homer was writing around 750 BC, so the story could be somewhat exaggerated.

At first the Mycenaean's seem to have been a bit shy of the Minoans who were pretty powerful and had a more sophisticated culture. But after the second disaster they arrived in Crete and it has been Greek ever since. Further migration of other Greek speaking people from the north took place gradually until they were established all over the Aegean islands. Apart from a common language, they shared a religion based on the Oracle of Apollo at Delphi. They would often meet up for festivals, which is how the Olympic Games first started in 776 BC.

For cities like Corinth and Miletus there was not enough fertile land and so they began colonising all along the Mediterranean coast, like the Phoenicians had done. In the east they colonised Byzantium and all along the Black Sea coast, and to the west as far as Spain. The French city of Marseilles was originally a Greek colony. There were other colonies in southern Italy and Sicily and one near Carthage in North Africa. All the colonies shared the same architecture, amphitheatres, art, literature, language and religion.

At first the city states were ruled by kings, but between 800 and 700 BC these were replaced by republics and the first democratic constitution was written in 507 BC. This is surely a major milestone in the evolution of humanity from the 'nothing' of the quantum vacuum to the present. It is also one which still proves to be too sophisticated and complex for many modern societies to handle.

Forty years before this the Persians invaded the whole of where Turkey is today and overran the Greek city states on the coast. When they rose up in revolt, Athens sent a fleet to their aid. It was in retaliation for this that the Persian king Darius sent a massive army to invade Greece. But the Persians were defeated at the Battle of Marathon in 490 BC. This was the start of the Persian Wars. It was in 480 BC that the legendary last stand of the 300 Spartans took place. They managed to hold up a Persian army of 300000 under Xerxes at a narrow pass called

Thermopylae. After three days they were eventually defeated by the Persians through trickery, and Athens was occupied. But later the same year in September they were beaten by a combined Greek fleet at the great naval battle of Salamis. The following year the Persians were finally defeated by the combined armies of Sparta, Athens, Corinth and other city states at the battle of Plataea.

In 448 BC a peace was agreed with the Persians and five years later Pericles becomes the leader of Athens, and its Golden Age begins. This is perhaps the best place to end our story of evolution from the very beginning. In Part Two we go on to discover how it was that we found our place in the universe.

Part Two

Discovering the Universe

Chapter Six
Escaping from Ignorance

Setting the Stage

Did you ever wonder who was the first person to realise that the moon shines because it is reflecting light from the sun? The story of Western astronomy probably starts with the Greek philosopher known as Thales of Miletus (625 – 550 BC). Miletus, as mentioned earlier, was a Greek city on the coast of modern Turkey. Thales studied in Egypt and is said to have used Babylonian calculations to make the first prediction of an eclipse of the sun in 585 BC. Because of this, he is given the credit for making the leap from superstition to the idea of the world being governed by physical laws, which is the basis of modern science.

What about places like Stonehenge? It's thought to have been built much earlier, roughly 3000 to 2000 BC. It is almost certainly some sort of astronomical site, particularly with certain alignments at the Summer solstice, but its main purpose was religious not scientific, and there is no written record. In China, sightings of Halley's Comet go back to 240 BC and perhaps even as far back as 1059 BC, but there is no evidence of making actual predictions. In July 1054 BC the Chinese recorded the first supernova ever seen. It took place in the constellation Taurus and was bright enough to be seen by daylight for nearly a month. They didn't know what it was, but modern telescopes found its remains in the shape of the Crab Nebula. You may have seen pictures of it looking like a giant explosion in space.

Ten years before Thales of Miletus died, Pythagoras of Samos (560-480 BC) was born. He was responsible for that theorem which most of us thought was pretty useless information when we were at school. Believe it or not though, it was a crucial step in the history of the world. The reason is that it was the first example of using mathematics to describe a physical law. That maybe doesn't seem very important, but without maths there would, for example, be no internet, smartphones, computers, satnav, engineering, economics, architecture, modern medicine, the list goes on and on. He was the originator of using the awesome power of mathematics to explain the world around us. He studied with the priests at Memphis in Egypt and at temples in Phoenicia.

Pythagoras is also important for another reason. He was the first to believe that mathematics is something independent of the human mind. Something that exists behind everything we see, but which is somehow outside both time and space. There will be much more about this later. Pythagoras founded a secret society that worshipped the principles of maths. They thought that numbers were what the universe is made of. This led to friction with the locals, and he and his followers came in for some heavy persecution. He wasn't averse to persecuting others himself though, like poor old Hipparsus who apparently let out the secret of irrational numbers. Numbers like pi that go on forever and ever as decimals but never come to an exact whole number. Irrational numbers thoroughly spoiled the Pythagoreans' whole religion about numbers being perfect. So, when Hipparsus let out the secret they had to kill him. The story goes that they threw him overboard after Pythagoras had strangled him!

Socrates was a few years younger than Pythagoras and lived in Athens. He didn't contribute much to astronomy, but he had a profound influence on the thinking of the Western world. He didn't actually write anything down himself about his ideas. What we know about him comes from his equally famous student Plato, who was born in 428 BC. One of the few certain facts about Socrates is that at seventy years of age he was tried by a court, condemned to death and executed in 399 BC. The reason for his execution was that his ideas, which have so influenced the world, were considered by his contemporaries in Athens to be too dangerous for the young.

The Death of Socrates

The death of Socrates is one of the most moving pieces of writing in the whole of classical literature. Plato gives an almost minute-by-minute account. The execution took place in Athens, and was carried out by drinking a chalice of poison, which took about twenty minutes to take effect. It began with numbness in the feet which spread slowly upwards to the heart. I still remember how I felt when my high school headmaster, who was an Oxford classics scholar, read it to a group of us students. He translated it directly from the ancient Greek. Needless to say, he considered it to be required reading for any civilised person!

Plato went on to write many works about the teaching of Socrates which have survived and which are still widely studied. Like the Pythagoreans, he also believed that mathematics is something that is independent of the human mind. Something 'nonphysical' which lies behind the appearance of our everyday world. He thought that everything in our ordinary world of human senses was a kind of inferior copy of this ideal world. People who believe this today are still called Platonists.

Plato's most famous student was Aristotle (384 – 322 BC), and it was Aristotle who produced the basis of all scientific thinking for the next eighteen centuries. He led an interesting life. His father was the court physician to the king of Macedonia in Northern Greece. When the king was assassinated, Aristotle moved to Athens where he studied under Plato. Later he returned to Macedonia in 343 BC, as personal tutor and advisor to the King's son Alexander, who would later become none other than Alexander the Great, who is still thought to be one of the greatest military commanders of all time. He conquered the mighty Persian Empire we talked about in the last chapter, and created one of his own that stretched from Northern Greece to India by the age of thirty-two, when he suddenly died in suspicious circumstances.

It's pretty certain that we have Aristotle to thank for Alexander's respect for knowledge and learning. In conquering the Mediterranean and most of Western Asia, he set up several centres of learning. The best known was in the city of Alexandria in Egypt where a famous university and library was built to house all of human knowledge. It was from Alexandria that the nations of Islam later took Greek knowledge and preserved it while Europe collapsed into the ignorance of the Dark Ages.

Classical learning only returned to Europe in the twelfth and thirteenth centuries through contact with Arab scholars in Spain.

Aristotle finally returned to Athens and set up a school called the Lyceum which was situated in his garden. His ideas about the sun and planets were wildly wrong, but unfortunately they were taken up by the early Christian leaders and became part of the doctrine of the Catholic Church which they were very reluctant to give up.

A Library Burns

Born only two years after Aristotle died, the first Greek philosopher to realise that the Earth and planets went around the sun was Aristarchos (320 – 250 BC) from the island of Samos. By looking at the size of the Earth's shadow during a lunar eclipse he came to realize that the sun had to be much bigger than the Earth, and very far away. This is what persuaded him that it was much more likely that the Earth went round the sun.

Considering the centuries of ignorance that followed, the history of astronomy might have been very different had not all but one of his manuscripts been burned in a fire that destroyed a large part of the great library in Alexandria. It happened accidentally about 50 BC during the Roman civil war, which ended in victory for Julius Caesar. After the war Caesar placed Cleopatra on the throne of Egypt. It may well have been as a result of one of the most sensational pieces of self-promotion ever recorded. The story goes that, in order to gain access to Caesar, she had her attendants roll her up in a large Persian rug meant as a peace-offering to the conqueror. When it was unrolled for him to examine, a beautiful young nubile princess emerged and stood before him, naked except for some very expensive jewellery. No wonder they became lovers, and that she later conquered the heart of Mark Anthony, another main player on the world stage at the time.

Thirty-three years younger than Aristarchos was Archimedes of Syracuse (287 – 212 BC). He was a Sicilian Greek mathematician who is not important for contributing to the story of astronomy, but because he is thought of as the first real scientist. He began what later became known as 'the scientific method'. He was the first person to devise experiments to test his theories, which he then expressed mathematically. A good example is the Archimedes Principle about the displacement of a body in water, and which he is famously said to have discovered in his bath.

The first human being to measure the size of our planet was another Greek, seventeen years younger than Archimedes. His name was Eratosthenes of Cyrene (270 – 190 BC). Educated in Athens, he later became the Director of the great Library in Alexandria. He discovered that on the longest day of the year, at a place in the south of Egypt called Cyrene, the sun did not cast any shadows at midday. He also noticed that sunlight was reflected directly back up a deep well, which meant that the sun had to be directly overhead. Using sticks placed upright in the sand, he measured the shadows cast on the same day in Alexandria which was much further north. He then measured the distance between Alexandria and Cyrene, and was able to work out the circumference of the Earth. His result was pretty accurate, which is quite remarkable. In later life he became blind and, because he could no longer read, he committed suicide by voluntary starvation. A choice of death for another famous figure in our story, who comes much later.

Twenty years after Eratoshenes, Hipparchus of Rhodes (170 – 125 BC) constructed one of the first extensive star catalogues. He chartered the positions of 850 stars in all, and came up with a scale for measuring their brightness which is still used today!

The Great Mistake

After Hipparchus there is a break of some two hundred years before we come to Ptolemy (90 – 170 AD). He was an Egyptian Greek astronomer who also lived in Alexandria. By this time Aristarchus' work had been lost in the fire, and Ptolemy was mostly influenced by the work of Aristotle. He made the great mistake of believing that the Earth stood still and that the sun went round it. He created what became known as the Ptolemaic system, with the Earth as the centre of the universe. He devised an elaborate set-up to explain the observed movement of the planets in the sky, which required eighty small circles or epicycles. Although it was cumbersome, it worked quite well with what astronomers could see, so people just accepted it. Along with Aristotle's ideas it was later incorporated into the Catholic Church as their official doctrine.

After this came the long decline of the Roman Empire, which officially ended in 467 AD with the humiliating abdication of the last emperor. With this event Western Europe was plunged into the Dark Ages, and all the knowledge of the Greek scholars was lost. Though it should be remembered that the empire in the East, with its capital at Constantinople (Istanbul today), went on successfully until the fourteenth century and gave rise to the Orthodox Church which includes most of Russia and Eastern Europe. It's easy to forget this when you are just thinking about Western Europe!

Nicholas Copernicus

A thousand years later, the ignorance of the Dark Ages slowly lifted with the arrival of the Renaissance, brought about by Christian contact with the Arab world. And so the knowledge of Greek and Roman civilisations arrived in Europe again. It arrived with the benefit of many years of Islamic scholarship in between, and spread quickly because of another hugely significant revolution. In 1450 Johannes Gutenberg invented the first printing press with movable type, which meant that books could be mass produced at reasonable cost. The technology was soon adopted widely in Europe.

The Sumerians had used cylinder seals made of metal and ceramic as long ago as 1500 BC, and the earliest Chinese woodblock prints date from around 220, but they did not have the advantage of mass production. Another big advantage for Europeans was the fact that their languages had a limited number of characters, just twenty-six letters, whereas the Chinese had upwards of a thousand.

One of the people who came across some of the rare references to the work of Aristarchos was Nicolas Copernicus (1473-1543). He was a Polish astronomer and clergyman who became a major figure in the story of science. He challenged the whole idea of an Earth centred universe, working out that the Earth must be revolving once a day which was why the sun appears to rise in the east and sink in the west as though it is going round the Earth. He came from Poland but studied in Italy until 1506.

He completed his major work, 'The Revolution of the Heavenly Spheres', in 1530, but did not publish it for thirteen years because he suspected it would cause an uproar within the church. It was the time of the Protestant Reformation in Europe, which was challenging the orthodoxy of the Catholic Church. The Holy Office of the Inquisition had been set up to root out all opposition to official doctrine. Some accounts say that Copernicus only saw his completed work published on the same day that he died, 24th May 1543. A friend of his thought it might be diplomatic to add a hasty preface explaining that the ideas expressed were: 'merely a computing device without prejudice to the Truth'. After others took up on his ideas, the church banned the book in 1616 and did not remove it from the index of forbidden books until 1835, 219 years later.

The Copernican system was much simpler than Ptolemy's, but because of its insistence on perfectly circular orbits for the planets, it was only slightly better in fitting the observations. However, as more and more observations were made, Ptolemy's old model had to undergo an increasing number of adjustments, and it began to look more and more contrived – even mildly absurd. So much so that when he was asked about it, King Alfonso of Spain is said to have delivered this cracking one-liner: "If I had been present at the Creation, I could have rendered profound advice!"[1]

The most important catalogue of stars made before the telescope was invented was the work of a Danish nobleman, Tycho Brahe (1546-1601), who was an amazing character. He lost a large part of his nose in a sword fight while duelling over an argument about maths! For the rest of his life he wore a false nose which he made himself out of silver. He was so impressed as a young man by a partial eclipse of the sun in 1560, that he decided to devote his life to astronomy. In 1572 he was the first person in Europe to sight a supernova which continued to shine brighter than Venus for a whole year.

It was an important discovery, because until then the stars were believed to be perfect unchanging 'eternal' religious objects. The King of Denmark, Frederick II, gave him an island for life where he built an observatory. Unfortunately, in 1596 Frederick's successor, Christian IV, forced him to leave his island. After three years wandering around, luckily without any further sword fights, he settled in Prague where he was given a castle by the mad Emperor of the Holy Roman Empire, Rudolph II. Tycho loved a good party with fine food and wine, and unfortunately he died prematurely two years later after a marathon drinking session.

Fortunately he had taken on as his assistant a German, Johannes Kepler (1570 – 1630), who was a follower of Copernicus, and when Tycho died Kepler inherited twenty years of the most accurate astronomical observations ever made. It was by using Tycho's charts that Kepler was able to work out the laws of planetary motion, discovering after a considerable struggle that the planets moved not in perfect circles but in oval ellipses. This made all the difference, so that the theory was now in close agreement with what was seen in the sky. But it was obviously a major blow for Catholic doctrine.

The Age of Intolerance

The first martyr to the Copernican view of the solar system was Giordano Bruno, an Italian who had been a Dominican monk. In his studies he came across the work of Copernicus and was convinced of its truth. He travelled widely, especially to Protestant England where he met with some of the leading political and scientific figures. He was actually the first person to think that the universe was immense, in fact so immense that there might be other alien civilisations. Of course the Catholic Church didn't like that. On his return to the continent he rather unwisely championed Copernican theory, so that the minute he set foot on Catholic territory in 1592 he was promptly arrested by Cardinal Bellarmine, the Pope's own theologian and a leading figure in the Counter-Reformation (the Catholic backlash against the Protestants). For seven years he tried to get Bruno to recant and admit that the sun went round the Earth. Bruno refused to the end, and in 1600 he was burned at the stake. It is said that he even refused strangulation before the fire was lit!

Galileo Galilei (1564-1642), is one of the greatest heroes of the Renaissance and one of the most important figures in the history of science. He was born in Pisa the same year as Shakespeare was born in England. He was the son of a musician, and so clever that he became a Professor of Maths at the age of twenty-five. He later moved to Florence where he lived with Marina Gambia but did not marry, despite having two daughters and a son, which was somewhat unusual then. They parted amicably some years later and he returned to Padua near Venice, but they still kept in touch. It was there in 1609 that he heard about a Dutch invention by Hans Lipershey called a spy-glass, which magnified things up to three times. In one night he figured out how it worked, and then made one with a twelve times magnification. It caused an immediate sensation. In a letter to Marina's brother he describes how he was, "called by the Signoria, to which I had to show it together with the entire Senate, to the infinite amazement of all; and there have been numerous gentlemen and senators who, though old, have more than once scaled the stairs of the highest campaniles in Venice to observe at sea sails and vessels so far away that, coming under full sail to port, two hours or more were required before they could be seen without my spy-glass."[2]

Galileo then constructed another telescope, stepping up the magnification to thirty times, and turned it on the night sky. It was a momentous event, and one which still gives me the shivers. This seemingly simple act changed forever our view of reality. It represented not only the birth of modern science, but removed beyond doubt the nature of our place amongst the stars. Between September 1609 and

March 1610 he wrote and published 'The Starry Messenger' recording, using watercolour illustrations, what his telescope revealed:

> Stars in myriads, which have never been seen before, and which surpass the old, previously known stars in number more than ten times.

> But that which will excite the greatest astonishment by far, and which indeed especially moved me to call the attention of all astronomers and philosophers, is this, namely, that I have discovered four planets, neither known nor observed by anyone of the astronomers before my time.[3]

He had discovered the moons of Jupiter. When he turned the telescope on our moon he discovered that its surface was not at all smooth but covered in craters and high mountains. On the day that 'The Starry Messenger' was published, Sir Henry Wotton, who was the English ambassador to the Doge's court in Venice, wrote to his superior in the Foreign Office in London.

> The mathematical professor at Padua hath…discovered four new planets rolling about the sphere of Jupiter, besides many unknown fixed stars; likewise…that the moon is not spherical, but endued with many prominences…The author runneth a fortune to be either exceeding famous or exceeding ridiculous. By the next ship your lordship shall receive from me one of the [optical] instruments, as it is bettered by this man.[4]

Galileo did become 'exceeding famous' and as a result of his observations was convinced of the Copernican view. In 1632 he published 'Dialogue on the Two Chief World Systems, Ptolemaic and Copernican', in which he attempted to be fairly diplomatic in his support for the Copernican system. I say 'fairly diplomatic', but it was written as a conversation between two people, one of whom represented the old views and the other the new. Unfortunately the person putting forth the old views was made to look a bit of a dolt, and some people thought it was a reference to the Pope!

The Trial

It was too late. Cardinal Bellarmine had been collecting evidence against Galileo for the previous fourteen years and, with the publication of the book, he made his move. Galileo was arrested and brought to trial before the Inquisition. He was 69 years of age. In the course of the trial he was shown the torture chamber so that he was in no doubt that it would be used. Details of the trial of Giordano Bruno had not been made public, but Galileo was informed of them. In the face of this opposition he was forced to recant. Everyone knows the story, but not many have read the actual document he had to sign. Here's the crucial bit:

> I have been pronounced by the Holy Office to be vehemently suspected of heresy, that is to say, of having held and believed that the sun is the centre of the world and immovable, and that the earth is not the centre and moves:
>
> Therefore, desiring to remove from the minds of your Eminences, and of all faithful Christians, this strong suspicion, reasonably conceived against me, with sincere heart and unfeigned faith I abjure, curse, and detest the aforesaid errors and heresies, and generally every other error and sect whatever contrary to the said Holy Church.[5]

He was sentenced to house arrest for the rest of his life. Amazingly, it was not until 1992, twenty-three years after Neil Armstrong stood on the moon, and three hundred and sixty years later, that the Catholic Church finally pardoned him.

Galileo continued to do science until his death in 1642. Another of his famous discoveries was that the speed at which objects fall, is independent of their weight. Or if you prefer, all objects fall at the same speed irrespective of how heavy they are. In other words, a lead shot put falls at the same speed as a golf ball, although it defies your senses if you feel how heavy both things are at the same time. Galileo worked this out by rolling cannon balls down an inclined chute, and probably not by dropping them from the leaning tower of Pisa as is popularly held. Because clocks were still inefficient, he used his own pulse to measure the time. In 1638 he published 'Discourses Concerning Two New Sciences' which established him as the father of the Modern Scientific Method. In order to arrive at the truth about the world you have to first make observations, then design and carry out experiments. These results must then be carefully and precisely described (preferably using maths), and then published so that others can verify the results by repeating the experiments. He also believed in the power of mathematics to describe the world. One of his best known quotes is his belief that, 'the language of Nature is written in mathematics'.

Apart from his discoveries, he was an able musician, artist and writer – a typical Renaissance man – and one of the greatest scientists that ever lived.

A Boy Playing on the Sea Shore

Sir Isaac Newton was born on Christmas Day 1642, the year Galileo died. It is the year that most people consider to be the beginning of the Modern Scientific era. Newton had a somewhat shaky start in life as he was born prematurely, and only three months later his father died. He inherited Woolsthorpe Manor in Lincolnshire, England, but was unfortunately left with his grandmother at a young age after his mother remarried. He went to the local grammar school in Grantham, and when he left there he first tried his hand at farming, a bit difficult to imagine considering what came later. But he soon gave it up as a bad job and went instead to Trinity College, Cambridge, in 1661.

Four years later, the university closed down because of the Great Plague which killed almost a third of the population of Europe. He went home to Woolsthorpe and spent his time studying. It proved to have enormous repercussions because it was here that he discovered calculus. If, like me, you never got that far at school, calculus is the basis of all modern physics. It allows you to work out the future position of a moving object, so that you can tell the path that will be followed by a ball thrown in the air, or a cannon ball, or even a planet going round the sun. It is a small comfort to know that actually everyone can do calculus without even knowing it. Each time you assess the speed of an approaching car before you step out across the road, your brain is using calculus to tell you when it is safe to cross.

For a long time Newton kept his discovery of calculus a secret. This led to a fierce argument later with Gottfried Leibniz, the famous German philosopher and mathematician, over who actually discovered it first. It was also at Woolsthorpe that Newton got the idea for gravity. He had a young niece as his housekeeper, and it is thought that the famous story about the apple hitting him on the head comes from her. It may well have been true. After he got the inspiration for gravity, he first applied the maths to the motion of the moon. Here is how he first recorded this huge event:

> I deduced that the forces which keep the planets in their orbs must be reciprocally as the squares of their distances from the centres about which they revolve; and thereby compared the force requisite to keep the moon in her orb with the force of gravity at the surface of the earth; and found them answer pretty nearly.[6]

More than just pretty. He was actually being rather modest because even this first calculation came out at 27¼ days. In discovering the inverse square law of gravitation he had unlocked one of the greatest secrets of Nature. The force of gravity is the same throughout the universe. It is still the means by which they work out each space mission. It doesn't change, so it's like a very basic bit of how the universe works. Scientists call it a 'constant' of Nature.

After the plague, Newton returned to Trinity in 1667 and was elected a Fellow. Two years later, Isaac Barrow resigned his Chair of Mathematics in favour of Newton, who was then only twenty-six. It was the same Lucasian Professorship later held by many famous scientists and recently by Stephen Hawking.

Having discovered such an important secret about the way the world works, it is strange to think that he never published it for twenty years. It was only after his friend Edmund Halley, (who discovered the famous comet) had actually pleaded, cajoled and harassed him into writing the book, and then even financing it himself, that it was published. 'The Mathematical Principles of Natural Philosophy', known as the 'Principia', appeared in 1687. It is considered by many to be the most important scientific book ever written. It contained Newton's three laws of motion as well as his law of universal gravitation, which he then used to explain the elliptical laws of planetary motion discovered by Kepler, including the movements of the moon, the Earth, and the ocean tides. It also explained Galileo's results on falling bodies.

The book was widely circulated in Europe, making Newton famous. A well known poet of the time, Alexander Pope, is often quoted for this couplet showing how Newton was regarded by the public:

Nature and Nature's laws lay hid in night:
God said, 'let Newton be!' and all was light.[7]

Despite his fame, Newton's only other work was in optics, in which he put forward his 'corpuscular' theory of light published in 1704, which proposed that light was actually made up of very small bits. His interest in science then took a bit of a dive, and he turned to theology in support of the Unitarians who didn't believe in the doctrine of the Trinity, (God the Father, God the Son and God the Holy Ghost).

Because of this disagreement, which was considered deadly serious at the time, he would have been forbidden promotion to Master of Trinity College, so he resigned his Lucasian Professorship and moved to London in 1696 to become Warden and then Master of the Royal Mint. He cleverly reformed the currency by preventing counterfeiting, and surprisingly it was for this – and not his science – that he was knighted by Queen Anne in 1705. In 1703 he was elected President of the Royal Society which he remained until his death in 1727. He also experimented a great deal with alchemy trying to turn ordinary metals into gold, but obviously without any success.

His intellect must have been quite prodigious. There is an anecdote that in 1696 the Swiss Mathematician Johann Bernoulli challenged his colleagues in Europe to solve a difficult maths problem. The time set for a solution was six months. Leibniz wrote back asking for a year. The challenge was given to Newton at 4.00pm on January 29th, 1697. Before leaving for work the next morning, Newton had discovered a new branch of maths and used it to solve the problem. He sent off the solution, requesting that it be published anonymously. The identity was apparently obvious. Bernoulli said – 'we recognise the lion by his claw'.

Despite his fame and achievements, he was clearly awed by the immensity of Nature. Towards the end of his life he wrote:

> I do not know what I may appear to the world; but to myself I seem to have been only like a boy playing on the sea-shore, and diverting myself in now and then finding a smoother pebble or a prettier shell than ordinary, while the great ocean of truth lay all undiscovered before me.[8]

References

1. Lerner, E.1992. *The Big Bang never happened.* London: Simon & Schuster. page100.
2. Bronowski, J., *The Ascent of Man,* B.B.C., London, 1973, page 202
3. ibid., page 204
4. ibid., page 204
5. ibid., page 216
6. ibid., page 223
7. Roberts, J.M., 1993. *History of the world.* London: BCA. page 544.
8. Bronowski, J. op. cit., page 236

Chapter Seven
The Universe Moves

Beyond our Solar System

Before Galileo everyone naturally thought that there was only the sun, the moon and planets, and the 'fixed' stars. That was it. That was all there was or had ever been. They even attached great religious and spiritual significance to what they saw in the sky. Imagine then, what a great shock it was to discover that there was much more to creation than could be seen with the naked eye, with our limited human senses. It was a huge expansion of human consciousness, and a revolution as big as any we've come across so far. For a start it showed that Catholic doctrine based on Aristotle was just plain wrong.

With increasing improvements in telescopes, astronomers began to notice small cloud-like objects, which they called 'nebulae'. Some even thought that they were spiral in shape, but the general opinion was that they were clouds of gas within the Milky Way, clouds that were perhaps condensing to make new stars.

In 1755, the famous German philosopher, Emmanuel Kant, published a book in which he made the bold suggestion that the nebulae were perhaps separate from the arc of the fixed stars. He called them 'island universes'.

The greatest practical astronomer of the day, and the man credited with pioneering the study of the universe beyond the Solar System, was Sir William Herschel. He constructed his own telescopes, the largest of which was forty feet or twelve metres long. A monster of a thing. With it, he discovered a new planet that he wanted to name George's star, after the reigning monarch at the time, George III. It ended up being called Uranus. But he also discovered both its satellite moons, as well as the moons of Saturn.

However, these pale into insignificance compared with his greatest discovery, often forgotten, which was a truly consciousness expanding realisation. You'll know what I mean when I tell you. By measuring the movement of the Sun against the position of seven stars in 1783, he discovered that the Sun, together with the planets, was actually moving through space! It must have been a frightening realisation.

It has now been confirmed that whilst you are reading this we are actually travelling at some 44 000 mph (70 000 kms) through space, along with the rest of the stars we can see in our bit of the galaxy, because the Milky Way, like almost every other galaxy, is actually revolving. For some strange reason I am always entertained by this remarkable piece of information. It's as though Nature is playing tricks on us. You have to wonder what Solomon and the Patriarchs would have made of it.

About Herschel; he was born in Germany in 1738, the son of a musician in the Hanoverian Guards. At the youthful age of nineteen he moved to England as a session musician (freelance), because he liked the place, having visited before with his father and the band. He must have been an accomplished musician because shortly after arriving in Bath he became the resident organist, and settled down, sending for his sister Caroline to join him from Germany. She helped him in making his telescopes and became a keen astronomer herself, discovering eight new comets.

After he discovered Uranus in 1781 he became the Court Astronomer, and was later elected a Fellow of the Royal Society. In 1802 he turned his telescope on the curious nebulae, making a catalogue of two thousand of them. In 1816 he was knighted, and in 1820 he published an even more extensive catalogue. Since most of the nebulae appeared to be a spiral shape, he supported Kant's idea that they were island universes. Unfortunately there was no way of measuring how far away they were. It took quite a long time before it was possible to do this.

Measuring the Milky Way

Henrietta Leavitt was a remarkable woman. She began as an amateur astronomer helping out at the Harvard University Observatory. Although she was deaf, she managed to get a permanent appointment on the observatory staff at a time when it was extremely difficult for any women to find work in the sciences. In 1912, while working on certain stars called Cepheid Variables which change their light output over a period of time, she discovered a method of measuring their distance based on their brightness. It was an important discovery still in use to day.

Then in 1914, a man called Vesto Slipher, working at the Lowell Observatory in Arizona, discovered another method of measuring distances using the Doppler Effect on light waves. This was something that went back to 1842 when an Austrian scientist, Christian Doppler performed an amusing and rather hair-raising experiment. He got an open railway carriage full of trumpeters playing as loud as they could, and sent it at some speed past a carefully chosen group of musicians who all had the hearing gift of perfect pitch. With this experiment he was able to show that the frequency of sound waves from a source moving towards a person will be increased, and waves from a source moving away will be stretched or decreased in frequency. This is just the familiar experience of hearing a police or ambulance siren growing louder as it approaches you, and then dying down as it gets further away.

What Vesto did was to apply this not to sound waves, but to light waves from the stars. Because of the star's distance, the light is shifted towards longer or redder wavelengths. The further away a star is, the greater the wavelength or distance between each wave crest, and so the greater the 'red shift'. He came up with a formula for measuring the distance of an object using its red shift.

Some years earlier an American newspaper crime reporter with a name straight out of a comic strip, decided to give up chasing crime stories and instead become an astronomer. His name was Harlow Shapley, and in 1915, using Henrietta Leavitt's method of Cepheid Variables, he was able to produce the first rough estimates of the size of the Milky Way. He showed that we live some 25 000 light years from the centre of the galaxy on the outer fringes of one of its spiral arms. It has since been found to be double this distance.

Resolving the Nebulae

The next character was George Ellery Hale. He was a successful astronomer, but one of his best talents was to persuade big American financial institutions to invest in the development of bigger and better telescopes. He managed to get the Carnegie Foundation to fund the construction of the 100 inch (2.5 metres) reflector on Mount Wilson near Pasadena in California – the greatest ever built at the time. Hale then appointed Edwin Hubble as chief astronomer at the new observatory – the man after whom the Hubble telescope is named.

Born in Marchfield, Missouri, the son of a lawyer, Hubble went to Chicago University, and then to Oxford University to study law on a Rhodes scholarship. Many other famous people have also been Rhodes Scholars, including President Clinton. They were endowed by Cecil Rhodes, the nineteenth century South African diamond magnate. His will allocated one of these scholarships each year to Bishops, the school I went to in Cape Town, South Africa. Surprisingly, the first stage for candidates was election by the whole student body, which I passed, but I failed at the next stage! I simply wasn't clever enough.

Perhaps unique amongst astronomers, Hubble was a renowned athlete and boxer, nearly becoming a professional heavyweight. He once fought an exhibition match against the champion Georges Carpentier. Later he changed from law to astronomy, and began his career in 1919 after serving with the U.S. Army as a major in France during World War One.

Using the new telescope, Hubble focused on the brightest of the nebulae known as Andromeda, and found that it was possible to resolve very faintly, individual stars in its outer spiral arms. Using Henrietta's formula, he worked out that Andromeda was around two million light years away.

Shapley had measured the Milky Way to be just 50 000 light years across. So at forty times the diameter of the Milky Way, there was absolutely no way Andromeda could be part of the Milky Way. And remember, Herschel had discovered two thousand of these things! Hubble had discovered another galaxy. The universe was suddenly far bigger than anyone had ever dreamed. Another big expansion to our consciousness had taken place. Kant had guessed correctly, the nebulae were island universes and our Milky Way was but one of them. It was a great shock. It is said that you could have heard a pin drop when the discovery was announced to a silent auditorium at the 33rd meeting of the American Astronomical Society in December 1924.

The Universe Expanding

Hubble set about measuring the distances to other nebulae. To do this, he had to use Slipher's red shift method. To help him, he enlisted the support of Milton Humason, one of the most unlikely characters in the history of astronomy, or science for that matter. Humason was a mule driver or 'skinner' who transported the observatory supplies five thousand feet (1 500 metres) up the mountain from Pasadena. "Humason would lead the column of mules on horseback, his white terrier standing just behind the saddle, its front paws on Humason's shoulders. He was a tobacco-chewing roustabout, a superb gambler and pool player and what was then called a 'ladies' man." [1]

He would spend the night on the mountain top before going back the following day. During his overnight stays he became curious about what was going on at the observatory. Hubble realised he was quick and good with his hands and so taught Humason how to measure red shifts. He was soon able to get high quality spectra from galaxies better than the professionals.

Whilst Humason photographed the spectra of the galaxies, Hubble began analysing the information. He noticed that, apart from the closest galaxies like Andromeda and the Magellanic Clouds, all the others were moving away from the Milky Way. Then in 1929, "while staring at the data, it dawned on Hubble that the red shifts of the galaxies were not random at all. There was a pattern: the further away a galaxy, the faster it seemed to be hurtling into the void."[2] Plotting a graph of speed against distance, Hubble found a straight line. He had discovered that the speed of the galaxies was proportional to their distance. In other words, the further away a galaxy was, the faster it was moving. This could mean only one thing – the universe itself was expanding! Stephen Hawking describes it as "one of the great intellectual revolutions of the twentieth century"[3].

To understand it properly we need a bit of explanation quickly. It won't take long. The kind of expansion the universe is experiencing is special because it allows shapes within it to stay the same. Unlike a hand grenade or a bomb, which explodes into hundreds of fragments, the universe is much more like a loaf of bread filled with raisins. Each raisin representing a galaxy. You mix the raisins into the dough and then put it in the oven. As the dough heats up it expands, taking the raisins outwards with it.

Most people think of the galaxies moving *through* space because they imagine that space is just there already. But what is actually happening is that space itself – like the dough – is expanding and taking the raisins or galaxies with it. So the galaxies are not expanding into empty space; they are being carried apart because space itself is expanding.

The idea that space is a thing – an actual substance in its own right – and that it can expand, is due to the greatest hero of modern science.

References

1. Sagan, Carl.1981 *Cosmos*. London: Macdonald & co. p 253.
2. Chown, Marcus.1993. *Afterglow of creation*. London: Arrow. Page15.
3. Hawking, Stephen. 1988. *A brief history of time*. London: Bantam Press. p 39.

Chapter Eight
A Remarkable Person

The Lover and Revolutionary

The word 'genius' is spread around pretty freely these days. It seems to be applied to anyone who stands out from the normal in any field you care to mention, from interior design and film directing, to fashion. In the past it was used much more carefully and selectively, and therefore carried more weight. In fact, there was only one widely recognised genius of the last century and he was Albert Einstein. The only other possible exception was in fact his great pal Kurt Gödel whom we will hear about later.

The truth about Albert Einstein is that he was actually a wild rebel and a revolutionary hero who single-handedly blasted human consciousness to a whole new level. To appreciate him you need to just forget the maths, as I am going to do, and think of him as a Ché Guevara of science. It's unfortunate that most of the photos of him were taken later in life, but when he was in combat gear, so to speak, and tearing down the walls of orthodox science, he was a young man in his twenties.

Figure 6 Albert Einstein.

Nor was he shy and reserved. In fact he was a regular Don Juan. You don't believe me? In July 2006 they released a whole bunch of his private letters which had never been seen before. There were many affairs and mistresses. He had an illegitimate daughter, Lieserl, who was adopted before he married her mother, Mileva. He later had two sons by Mileva, but while still married in 1912 he had a relationship with his cousin, Elsa, who he then married after divorcing Mileva in 1919. Four years later he took another lover, Bette Neuman. There were many others. One was a beautiful Berlin socialite, Ethel Michanowski, who followed him to Oxford where he was lecturing, only to discover that he was involved with a third woman there! Another lover was a wealthy heiress, Margarete Lenbach, who used to send round her chauffeur-driven car to collect him for their late night liaisons! One of his longest love affairs took place during the Second World War with a real Soviet spy called Margarita Konenkova. It was only long after she returned to Russia in 1945 that the truth came out. Albert didn't know she was a spy of course, but then neither did the FBI who were actually watching *him*!

People also forget that he actually took part in the real live revolution that ended the First World War in 1918. Einstein was already a famous professor in Berlin at the time, with a reputation for being left-wing. So when the German navy mutinied on 4th November, the news spread to Berlin and the people poured onto the streets and joined the soldiers in erecting barricades. There was so much chaos that the Kaiser abdicated on the afternoon of the 9th November and fled into exile.

The revolutionaries immediately began arresting members of his government, including the rector and deans of the university who they then threw into prison. While chaos reigned on the streets, several of the professors contacted Einstein and asked him to try and get them released. With two chosen friends and a prepared speech in his pocket, they made their way to the Reichstag building. Luckily, when they approached the crowd of guards outside, one of them recognised him as an 'Obersozi' (top-rank Red or 'lefty'), and they were let in. He spoke to a revolutionary council that was gathered there, and urged them not to loose sight of the freedom they had gained by replacing it with a dictatorship of the proletariat. It was greeted with silence! They were then told that they would have to get permission from Friedrich Ebert, the newly proclaimed president of the Reich.

They had to go through the streets again to the Reich Chancellery where they met with the new president who recognised Einstein and agreed to sign the release for his colleagues. In letters to friends Einstein was delighted by the revolution because it swept away the old order of the ruling classes. He described his part as though it was an exciting adventure. That day he was supposed to give a lecture and he wrote in the register: "No class because of revolution"! So you see, it is wrong to think of Einstein as just a venerable old man with untidy white hair.

A Slow Starter

Compared with the average, Einstein was definitely a late developer. He couldn't talk until he was three years old, by which time Mozart had probably composed a symphony. His whole school career was unremarkable, except that he was actually expelled from high school at the age of fifteen for being a 'disruptive influence', which is a polite way to describe a teenage delinquent.

Not a good start, but it gets worse. In 1895 he applied for admission to Zurich Polytechnic and failed the entrance exam! Of all subjects, his maths was not good enough. He spent a year cramming and passed at the second attempt. Again, he only just scraped through and, because of his poor results, he was not able to get a job as a teacher. This meant that he had to fall back on teaching students privately, which was very poorly paid.

Finally in 1902 he landed a job at the Swiss Patent Office in Berne, as a patent inspector, junior grade. This was a steady job which allowed him to get married and become a Swiss citizen. He may well have settled down and never been heard of, except that he found the work easy, which gave him ample spare time to think about things.

Three years later, in 1905, and without even so much as a post graduate qualification, he revolutionised overnight our most fundamental ideas about Nature, and changed forever the way we look at the world.

Space and Time

He published three papers in one edition of the journal 'Annalen der Physik'. One was on something called the statistics of Brownian Motion, the second was a paper on the photoelectric effect, which proved that light did not just come in waves but also as separate particles or photons, and the third was Special Relativity. Any one of them could have earned him a Nobel Prize. It was the second that did, but only seventeen years later.

The paper on Special Relativity was thirty pages long and had the seemingly harmless title: 'On the Electrodynamics of Moving Bodies'. It was anything but harmless. In our ordinary common sense view of the world, there can hardly be anything more basic than our feeling of space and the passage of time. Yet the Special Theory proves that these two completely separate things are really one, now called space-time. Hawking says, "we must accept that time is not completely separate from and independent of space, but is combined with it to form an object called space-time"[1].

One of the most interesting bits of the theory is about the nature of light. According to Newton's laws of motion, and our own common sense, if one sports car goes at a hundred miles (160 kms) per hour, and another one goes at a hundred and fifty (240kms), then on a straight track the second car will catch up and overtake the first one. This is not the case with light. Nothing can catch light. Its speed is the same everywhere, no matter how fast you are travelling and no matter from where.

The Mystery of Light

Paul Davies, who is an internationally renown scientist and author says Einstein came, "to a remarkable – indeed scarcely believable – conclusion: the speed of light is the same everywhere for everybody, and this is true no matter how they are moving."[2]

Light is the only thing we have discovered in the universe which has no mass or weight, and therefore it travels at what seems to be the ultimate cosmic speed limit which we measure to be 186 000 miles or 300 000 kilometres per *second*. That's roughly four times right round the world in *one* second. That's fast. Here's another example. Just click your fingers. In the time it took to do that, a ray of light could have made the round-trip between Europe and America about thirty times.

Why light travels at this particular speed and not any faster or slower, no one yet knows. It's like a sort of very basic boundary to the whole way the universe is constructed. And, like gravity, it is called a 'constant' of Nature. Perhaps the greatest quest of all time is for science to find out why these constants have the exact values they do have. This is the whole object and focus of the massive investment that has been made over the years to find the Theory of Everything that will explain why these constants are what they are.

Light is fast, but this is only the beginning of the mystery. Another consequence of Special Relativity is that 'energy' and 'mass' (all matter) are the same thing in different forms. This is the world's most famous equation and the only one I'm going to mention:

$$E = MC^2$$

Leon Lederman, a well known Nobel Prize winner, once explained it like this, "mass M and energy E are one and the same thing, just measured in different units, and to correct for those units we must use the maximum velocity of the speed of light C."[3] So anything and everything that possesses any weight at all can be turned into pure energy – in fact is the same thing as light! It sounds absolutely preposterous, but forty years later the proof of this equation was demonstrated with the detonation of the first atomic bomb and the start of the nuclear age.

The First Infinite

But it also means that nothing in the universe which possesses mass, including ourselves, can ever reach the speed of light. Something much more mysterious begins to happen as we approach the speed limit. Listen to this from Hawking: "Because of the equivalence of energy and mass, the energy which an object has due to its motion will add to its mass. In other words, it will make it harder to increase its speed…as an object approaches the speed of light, its mass rises ever more quickly, so it takes more and more energy to speed it up further. It can in fact never reach the speed of light, because by then its mass would have become infinite, and by the equivalence of mass and energy, it would have taken an infinite amount of energy to get it there"[4].

As we speak, the biggest machine ever built, the Large Hadron Collider at CERN near Geneva, can accelerate tiny bits of atoms to 99.999% of the speed of light. But they can never actually get there because it would take more than all the energy in the universe to arrive. In other words, it's impossible to reach the speed of light even using the smallest and lightest objects we know of, because to do so would take an infinite amount of energy. Is this some kind of magic? Makes you wonder what could possibly have designed that?

This is the first time we have come across a real live infinite, so let's remember it, because later we'll be discovering all sorts of wonderful things about infinities. A lot of people are scared of the very sound of 'infinities', but we will become quite friendly with them as we go along. The reason I want to look at them is that they are without any doubt, the most awesome objects known to us. One of the best known infinities resides in the heart of a black hole, and there is thought to be at least one in every galaxy. But more about this later.

So nothing in the universe can ever reach the speed of light because its mass would become infinite, and it would need an infinite amount of energy to get it there.

Also, because of Special Relativity, we find that time slows down as you approach the speed of light. Despite how weird this may seem, the correctness of the theory is proved every day of the year in the same experiments with atomic particles in big accelerators. Because of the speed they are travelling at in these experiments, time slows down for the particles relative to the observers or targets. When at rest the particles would only exist for billionths of a second, and they wouldn't be able to measure them, but because of the speed at which they are travelling, they live much longer. The same thing, known as time dilation, is happening every second above our heads in the upper atmosphere, where cosmic ray particles going at very high speeds collide with particles in the air producing much longer lasting interactions. This allows the interactions to be recorded in high altitude balloon experiments. The other bit of the theory that still shocks me is that if you accelerated an ordinary school ruler to close to the speed of light it would actually get shorter and shorter and would disappear if it could ever reach the speed of light!

The Twin Paradox

Another bizarre consequence of Special Relativity is the famous Twin-Paradox. Take two identical twins. If one twin were to leave Earth in a spaceship travelling at close to the speed of light on a return trip to Alpha Centuri, (our nearest star) which is four light years away, then by the time he or she got back they would be eight years older (four years there and four years back). But the twin who stayed behind on Earth would then be not eight but sixteen years older! And they wouldn't be twins any longer. (Or would they?) This aspect of the theory was used to great effect in the "Back to the Future" movies.

Light is one of the most mysterious things we know of, and also one of the most important keys to how our reality works. Richard Feynman, another Nobel winner and one of the most celebrated scientists of the 20th century, pointed out that because a photon of light is already travelling at the speed of light, it can have no experience of time. For it, time just does not exist. Just as it has no mass, it also experiences no time. And if it experiences no time then it can't be aware of any distance. This is how strange light is. I can't help wondering what the folk of the Old Testament would have thought about it.

General Relativity

After the publication of his three revolutionary papers in 1905, Einstein received recognition for his work. In 1909 he obtained a junior professorship at the University of Zürich, followed by a full Professorship in Prague in 1910, and then a full Professorship back at Zürich in 1912. In 1913 he obtained one of the most prestigious posts in Europe, as Director of the Institute of Physics at the Kaiser Wilhelm Institute in Berlin where he had taken part in the revolution.

During these years up until 1915, he followed up his work on Relativity. Brilliant as it was, the Special Theory contained a major problem in that it didn't account for the action of gravity. Newton had just assumed that gravity was a God-given force. It acted between all bodies according to the inverse square law. However, it did concern Newton that there was no explanation as to why this should happen. He worried that it was a kind of instant action-at-a-distance that didn't have any explanation.

In thinking about gravity, Einstein came to the conclusion that the action of gravity was perhaps the same as the action of acceleration, which you feel pushing you back in your seat as an aircraft takes off. It feels as though it might be the same sort of force as gravity. Because there were no jet aircraft in his day, he thought about an elevator. If the steel cable in the elevator shaft suddenly snapped, the people inside the elevator would become weightless as they fell. In other words, they would not experience the sensation of gravity. Therefore gravity was perhaps just a form of acceleration.

He knew that gravity acted instantly between large bodies like planets and stars, but this actually contradicted his Special Theory of Relativity which says that nothing can travel faster than the speed of light. Steven Weinberg says the solution was "a profound analogy between the role of gravitation in physics and that of curvature in geometry...Einstein leaped to the conclusion that gravitation is nothing more or less than an effect of the curvature of space and time."[5]

THE FINAL MYSTERY

The most frequently used example to explain it goes like this. Think of a billiard table, the top of which is a tightly stretched sheet of rubber instead of slabs of slate covered with green cloth. If you roll a marble across the sheet it will go in a straight line. But if you get a heavy object like a bowling ball and place it in the middle of the rubber sheet it will cause a large depression. Now, any marble you roll across the sheet will be affected by the depression. Instead of going straight it will follow a curved path around the depression in the rubber sheet. That is where the so called 'force' of gravity comes from. There really isn't any force. Objects are simply following the path of least resistance, through a curved portion of space, (or space-time). The object can be a marble, a planet, or a person.

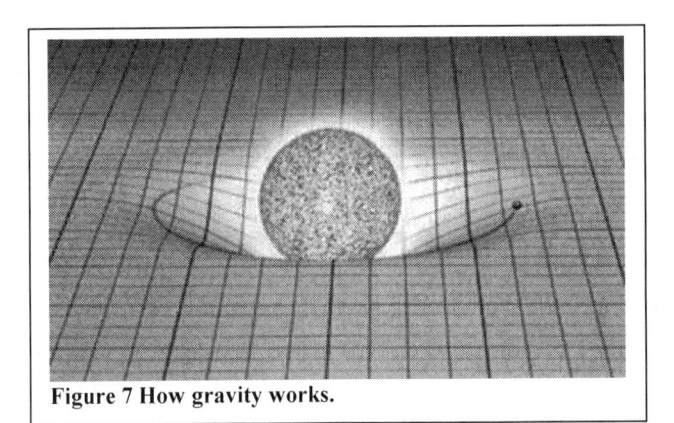

Figure 7 How gravity works.

The sun, being a large object, causes a big depression in the fabric of space-time, and the Earth and the other planets are like marbles going round the depression created by the Sun. They don't actually fall into the sun, as a marble would eventually, because the bowling ball or sun is completely surrounded by curved space, not just on its underneath side, so the speed it's travelling or falling at, exactly matches the curvature and stops it crashing into the bigger object.

To come up with General Relativity, Einstein had to learn about the maths of curved surfaces. It's hard to believe, but he actually thought of himself as bad at mathematics, so he asked his friend Marcel Grossman from his days at the Patent Office to teach it to him.

A Triumph for Mathematics

About a hundred years before this, mathematicians had found a new kind of geometry that was different from Euclid's that we all learn at school. It might interest you to know that Euclid was a contemporary of Aristarchus who lived in Alexandria around 300 BC. I can't escape the thought that they probably met at the great library most days for a chat, and maybe even had a few drinks together at a local place of refreshment. He wrote thirteen books called 'The Elements of Geometry' which brought all the theorems of geometry together, proving each theorem on the basis of the one before, except for a well-known problem called the Fifth Postulate which stumped him.

For two thousand years no one thought that any other kind of geometry existed, until Georg Riemann began trying to solve the Fifth Postulate. In doing so he discovered a whole new field of geometry. It was this geometry that Marcel taught to Einstein which he then used to come up with the General Theory. It is by far his greatest creation and the one that made him world famous.

It provided a totally new and much deeper explanation of what gravity really is. There was no more unexplained action-at-a-distance. John Wheeler, who was another major figure in the physics of the 20th century, described it like this: "space tells matter how to move and matter tells space how to curve." General Relativity is widely thought to be one of the greatest triumphs of the human mind. Not only that, General Relativity is also said to be a beautiful theory. It is said to have a kind of inevitable logical correctness. Einstein said, "the chief attraction of the theory lies in its logical completeness. If a single one of the conclusions drawn from it proves wrong, it must be given up: to modify it without destroying the whole structure seems to be impossible."[6] In a postcard to Arnold Sommerfield, an older colleague, Einstein said, "Of the general theory of relativity you will be convinced, once you have studied it. Therefore I am not going to defend it with a single word."[7]

Who is the third person?

Having mentioned the beauty of the theory, it is as well to remind ourselves of a popular anecdote which Hawking retells. In the early 1920's a journalist told Sir Arthur Eddington, a well known British astronomer, that he had heard there were only three people in the world who understood General Relativity. "Eddington is said to have paused, and then replied, 'I am trying to think who the third person is.'"[8]

Whilst hundreds now know it back to front, and many have extended it enormously, it will sadly remain a closed book for most of us, including myself. I personally don't think it matters too much so long as we can grasp its significance. For me, the deepest mystery by far is why mathematics should describe the real world at all. In talking about Riemann, Weinberg admits, "It is very strange that mathematicians are led by their sense of mathematical beauty to develop formal structures that physicists only later find useful, even where the mathematician had no such goal in mind."[9] If Steven Weinberg thinks it 'very strange', then I am blown away by it.

Earlier he makes the point that, "the new geometry was developed in order to settle a historic question about the foundations of geometry, not at all because anyone thought it applied to the real world."[10] That is the formidable power of mathematics.

Mathematics seems almost to come from a level which is deeper than physical reality. The truth behind an equation exists long before it is discovered. In fact, the truth of it almost certainly existed before time began. General Relativity did not contradict Newton's laws, it merely replaced them. It showed that Newton's work was just an approximation of something much bigger. In fact, General Relativity is in total agreement with Newton, but it extends Newton's idea of gravity much further by explaining the action-at-a-distance which Newton's equations could not.

The sheer wonder of the power of mathematics to describe our reality is made clear in this next anecdote. In 1952 Einstein was asked to be the first president of the new state of Israel, but he declined, saying that he was too naïve for politics. Hawking thinks there may have been another reason, and quotes Einstein himself:

> Equations are more important to me,
> because politics is for the present,
> but an equation is something for eternity.[11]

Beat that if you can!

General Relativity Goes Public

It took Einstein ten years to complete the General Theory of Relativity. It was first announced to the Berlin Academy of Sciences towards the end of 1915, and published in 'Annale de Physik'. The paper was called 'The Foundations of the General Theory of Relativity'. To begin with, the theory was only known to academics in Germany. International communication was made difficult because the First World War was raging in Europe.

The first test of Einstein's Theory he carried out himself. It was about the orbit of the planet Mercury. Since 1859, when telescopes were improving, astronomers had found from observation that Mercury's path around the sun did not agree precisely with Newton's laws. It actually swings slightly, so that each year the orbit is a bit different from the last. Einstein applied his new theory to Mercury's orbit and found that it worked perfectly – proof that the theory actually described Nature to a much greater accuracy than Newton. He admitted that he was beside himself with delight for several days after this discovery.

Stephen Hawking reminds us that an ordinary every day proof of General Relativity is now built into most family cars. To someone floating high up above the Earth it would seem as though everything down below was taking longer to happen. This is exactly the case with satellite navigation systems. "If one ignored the predictions of general relativity, the position that one calculated would be wrong by several miles!"[12]

The mathematics of Einstein's theory also predicted something that no human had ever thought possible – the fact that light would be bent by gravity. In 1916 he showed that the light coming from stars far behind the Sun, which passed close to the sun's edge, would be bent inwards. In other words, any starlight that just grazed the edge of the sun would be affected by its gravity. However, the only way to see if this was true would be to measure the starlight during a total eclipse of the sun.

Sir Arthur Eddington, whom I mentioned earlier, heard about Einstein's prediction through William de Sitter, who lived in Holland, which was neutral territory at the time. As luck would have it, a total eclipse was due in the tropics on May 29th 1919. Hastily, Eddington organised two expeditions, one to Brazil and another to the island of Principe off the west coast of Africa, to test the theory. General Relativity was proved true by experiment! The results were announced to a special combined meeting of the Royal Astronomical Society and the Royal Society on 6th November 1919. They received huge press coverage all over the world, and Einstein became the most famous scientist of the century.

Einstein has been called the father of Cosmology, and for good reason. Cosmology is the science of understanding the whole universe, and General Relativity is a theory that is capable of describing whole universes. John Gribbin at the University of Sussex and a well known science writer explains, "The important point is that Einstein did not have to expand his theory, in some sense, to make it capable of dealing with the whole universe. General Relativity, from its birth, dealt with whole universes quite happily."[13]

His Biggest Blunder

In 1917 Einstein made his first attempt to describe a simple universe using his new theory. The results seem to have baffled him. The equations told him that the universe must be either expanding or contracting. According to the maths it couldn't sit still. But that seemed just crazy. At the time no one ever thought that the universe was anything but static, Einstein included. The only way he could get his model of the universe to stand still was by adding an extra term to the equations, which became known as the 'Cosmological Constant'. Stephen Hawking says, "Einstein introduced a new "antigravity" force, which, unlike other forces, did not come from any particular source, but was built into the very fabric of space-time."[14] This was to become much more important long after he died, but he didn't know that. At the time he just made it up. In other words, he sort of cheated, although he told everyone why he was doing it of course.

Einstein ends the paper by saying about the bit which he added: "That term is necessary only for the purpose of making possible a quasi-static distribution of matter, as required by the fact of the small velocities of the stars."[15]

What must he have felt, when in 1929 Hubble announced to the world that the universe *was* expanding! He later referred to it as the biggest blunder of his life. Had he believed the equations of General Relativity, he would have predicted one of the greatest scientific discoveries of all time.

This is one of the most spectacular examples of the eerie power of mathematics to describe the real world. It still makes the hair at the back of my neck stand on end.
I will let Steven Weinberg have the last word:

> This is often the way it is in Physics – our mistake is not that we take our theories too seriously, but that we do not take them seriously enough. It is always hard to realize that these numbers and equations we play with at our desks have something to do with the real world.[16]

References

1. Hawking, Stephen W., 1988. *A brief history of time.* London: Bantam Press. p 23.
2. Davies, Paul., 1990. *Other worlds.* London: Penguin. p 37.

3. Lederman, Leon., and Shramm, David., 1989. *From quarks to the cosmos.* NewYork: Scientific American Library. p 43.
4. Hawking, S., op. cit. p 21.
5. Weinberg, Steven. 1993. *Dreams of a final theory.* London: Vintage. p 79.
6. ibid., p 107.
7. ibid., p 80.
8. Hawking, S., op. cit. p 83.
9. Weinberg, S., op. cit. p 125.
10. ibid., p 122.
11. Hawking, S., op. cit. p 178.
12. ibid., p 33.
13. Gribbin, John. 1986. *In search of the big bang.* London: Corgi Books. p 90.
14. Hawking, S. op. cit. p 40.
15. Gribbin, J., op. cit. p 91
16. Weinberg, Steven., 1987. *The first three minutes.* London: Fontana. p 128.

Chapter Nine
Evidence of Creation

Why the Stars Haven't Shone Forever

It was a great shock to suddenly realise that the reality we live in is actually undergoing enormous change on a gigantic scale. But this was nothing compared to the realisation that came next. If the universe is expanding now, then it must follow that in the past all the galaxies were much closer together. Even further back in time, perhaps the whole universe was contracted to a single beginning.

Nowadays everyone grows up knowing that there was a beginning to the universe. But I can still remember when I first heard about it. It hit me like a thunderbolt. It seemed just wild, I mean human beings weren't meant to know things like that. Everyone got by on Genesis even though you thought it was just a myth.

One of the first people to think the universe didn't just go on and on forever was the German philosopher Heinrich Oblers. His observation is known as Oblers' Paradox. It's so simple it is wonderful, and gives us a useful insight into what an infinity might really be like. In 1823 he argued that if the universe was infinite, then so much starlight would have gathered over such a long time that it would never be night. In an infinite static universe, nearly every line of sight would end on the surface of a star so that the sky would be a solid mass of stars and it would never be dark at night.

Oblers' answer to this was to say that the light from distant stars was somehow absorbed by clouds of gas that got in the way. But this doesn't work either, as Stephen Hawking explains: "if that happened the intervening matter would eventually heat up until it glowed as brightly as the stars."[1]

The only way out of this trap is to assume that the stars have not been shining forever, and therefore the universe is perhaps not infinite.

Different Beginnings

One of the first people to consider an actual physical beginning to the universe was the priest and astronomer Georges-Henri Lemaitre, who won medals for bravery in the 1st World War. Using Einstein's equations he came up with an initial kind of primal explosion. When he showed Einstein his ideas at the Fifth Solvay Conference in Brussels in 1927, Einstein is said to have commented that his calculations were correct but his physical insight was abominable! His former teacher Eddington was not more sympathetic either, and was actually offended by talk of cosmic beginnings: "it seems to me that the most satisfactory theory would be one which made the beginning not too unaesthetically abrupt."[2] So you can see why I was shocked. Even the brightest scientists found it pretty difficult to handle the idea of an actual 'Beginning'.

Not surprisingly, when Hubble came up with the evidence not many people were quick to take up Lemaître's exploding atom. A different explanation was offered by Sir Fred Hoyle, whom we mentioned earlier, and his colleagues at Cambridge, Sir Hermann Bondi and Thomas Gold, all of whom worked together on the development of radar in the Second World War. Their idea was known as the 'Steady State' theory, which they based on something they called the Perfect Cosmological Principle. In this model, matter, in the form of the simplest atom of hydrogen, was being continually created between the galaxies as they moved apart. In this way, new galaxies could slowly form and the universe was without an abrupt beginning.

Then in 1931 George Gamow, a Russian scientist from Petrograd, disembarked in New York, immigrating to the U.S. He had worked with famous names like Rutherford in Cambridge, and Niels Bohr in Copenhagen whom we'll hear about later.

He was the first person to take the idea of a big bang seriously because he wanted to know how all the elements like hydrogen, oxygen, carbon and iron had been formed in the first place. Atoms of the ninety two elements in Nature make up everything in the universe (except dark matter), from cabbages to kings, and mountains to merry-go-rounds, but no-one knew how they had been made originally. Gamow's guess was that they were created in the initial explosion that gave birth to the universe. He was a theoretical physicist, and one of the first people to realise that the world of the very small (the world of atoms) could explain the world of the very large (the behaviour of the cosmos).

Unfortunately, the calculations showed that only the two lightest and simplest kinds of atoms, hydrogen and helium, could be formed in this way, and hence that the early universe could not have made all the heavier elements. Astronomers knew from analysing starlight that most of the universe of stars and gas we can see, is made up of roughly 75% hydrogen and 25% helium. Fred Hoyle and his colleagues would later discover that all the heavier elements like carbon, sodium and iron were formed in the hot interiors of stars.

Meanwhile, Gamow published his ideas about the universe being born in a big explosion, and the two rival theories battled it out for ten years. It was only later, in 1958, when Fred Hoyle was being interviewed on BBC Radio, that he coined the phrase 'Big Bang' about his opponents' theory. He was making fun of it, suggesting that it was much too simple an explanation – but the name stuck.

A new kind of telescope then entered the picture. Normal optical telescopes can only collect visible light – that narrow band of the electromagnetic spectrum which we see with our eyes. The new telescopes were able to collect radio waves, which meant that astronomers could look much deeper into space, and therefore further back in time.

Sir Martin Ryle at the University of Cambridge, who also worked on radar, became interested in radio astronomy and built one of the first radio telescopes. With it he mapped the distribution of radio galaxies across the northern sky, and found that they were not distributed evenly across space, as they should have been according to Hoyle's Perfect

Cosmological Principle. Gamow's wife, Barbara, was so delighted with the news that she wrote this little ditty:

"Your years of toil"
Said Ryle to Hoyle,
"Are wasted years, believe me.
The steady state
Is out of date.
Unless my eyes deceive me,
My telescope
Has dashed your hope;
Your tenets are refuted.
Let me be terse:
Our universe
Grows daily more diluted!"
Said Hoyle, "You quote
Lemaître, I note,
And Gamow. Well forget them!
That errant gang
And their Big Bang –
Why aid them, and abet them?
You see, my friend,
It has no end
And there was no beginning,
As Bondi, Gold,
And I will hold
Until our hair is thinning."[3]

The Embers of Creation

In 1946 Gamow took on a research student, Ralph Alpher, who was then joined by a Princeton graduate, Robert Herman. Gamow set them to work on the problem of what matter would have been like just after the big bang. They realised that the universe must have begun at a temperature of perhaps billions of degrees centigrade, and that it would gradually have cooled as time passed and space expanded.

It was whilst doing these calculations that a spine-tingling thought dawned upon the two young researchers. What if some of the heat from this 'first event' was still left over in the universe today? They set about estimating what this temperature might be. The figure they came up with was –268°C, which is just five degrees above absolute zero. Absolute zero is the lowest possible temperature that can be attained by anything. It has a special function in physics. Marcus Chown, a well known science writer, explains, "When an object is cooled, its atoms move more and more sluggishly. Absolute zero (which on the Celsius scale is equal to -273.15°C) is the temperature at which they stop moving altogether."[4] Gamow's team published their prediction, but at the time there was nobody who knew of any technology that was capable of measuring such a faint glimmer from the beginning of time. Although they had published their results in 'Nature', one of the leading science journals in the world, as often happens, and despite being such a vital piece of knowledge, the prediction was forgotten by science for almost twenty years.

Then in the early 1960's, Robert Dicke, a physicist at Princeton University, not far from New York, began thinking about the Big Bang. He had also worked on radar, but in the U.S., at the famous Massachusetts Institute of Technology. He wanted to know what happened before the Big Bang. Ninety-nine out of a hundred scientists would have said that such a question was not within the realm of science; it was a subject for religion. But not Dicke. He came up with what is called the oscillating or 'bouncing' universe theory.

His idea was that if the universe is expanding now from an explosion in the past, then eventually it will slow down and stop expanding in the future. At this point the force of gravity should take over, and gradually begin to pull all the galaxies back together again into a 'big crunch'. When this 'crunch' got very compressed then perhaps it would explode once more and another universe would begin all over again – thereby providing an explanation of eternity!

For this to happen he realised that there would have to be really massive temperatures at the Big Bang/crunch moment. It set him wondering if some of this heat might still be detectable. He had never heard of Gamow's work at the time.

His suspicion was that this faint ember from the beginning of time would by now be barely glowing, and would therefore appear as short wavelength radio waves. He had two students working with him, so he set them the task of building a special kind of antenna or receiver that might be able to pick up this background radiation, which they knew would have to be coming from every part of the sky. Meanwhile, another colleague, Jim Peebles, did the calculations to see what the temperature of the radiation might be. At his first attempt he came up with a figure about ten degrees above absolute zero. He didn't realise that others had come up with a similar result all those years earlier.

A Lucky Accident

In the summer of 1964, Dicke and his assistants were building the antenna for their detector on the roof of a lab at Princeton. Unbeknown to them, only a few kilometres away at the Bell Telephone Company's laboratory in New Jersey, two scientists were working away busily with sleeves rolled up, scrubbing pigeon droppings off the aluminium surface of a huge horn shaped telescope that the company had built to track Echo One and Telstar, the first ever communications satellites.

Their names were Arno Penzias and Robert Wilson. They were both radio astronomers employed by Bell Labs to push the technology of radio waves to its limits. They thought this might best be achieved through experiments in astronomy, and planned to examine the faint radiation coming from the halo of matter that surrounds the Milky Way. It would require some very fine tuning, and the main problem would be to isolate the signals they wanted from all the other interference that existed in their equipment, and in the surrounding atmosphere.

Figure 8 Horn telescope at Holmdel

Finally, when they thought they had eliminated all possible sources of interference, they found that their horn telescope was still generating more radio static than they expected. Thinking that it might be a man-made signal, they pointed it away from the direction of New York, but the ghostly static persisted. They then tried pointing the horn right around the horizon in every direction, but the 'excess noise' was coming from the whole sky.

Some pigeons had been roosting in the twenty-foot (7 metre) horn and left their tell-tale droppings on the metallic surface. That's why Penzias and Wilson were cleaning it up: just in case it was producing the unwanted signal. They had carefully trapped the pigeons earlier, and set them free by company mail to another Bell lab forty miles (64 kms) away. By the time they had finished cleaning up the mess two days later, the pigeons were back. This time the scientists resorted to a more terminal solution.

Still the interfering hiss remained. As a last ditch hope they covered all the rivets in the horn with aluminium tape in case they were causing the problem. It didn't work. By now it was April 1965. One day, while phoning a fellow radio astronomer, Arno mentioned their frustration with the interference. The friend knew another friend who had mentioned to him Robert Dicke's research at Princeton, and he suggested that Arno phone him up.

He got in touch with Dicke during a lunch break, at which the Princeton team were having an informal meeting. Dicke was on the phone for some time, and from the questions he was asking and the information he was getting, the others knew that something was up. When Dicke came off the phone he said, "Well boys, we've been scooped."[5]

The temperature of the universe provided by the microwave background radiation which Penzias and Wilson had discovered was 3.5 degrees above absolute zero. It was almost certainly the leftovers of the act of creation.

It was the biggest discovery since Hubble had found that the universe was expanding. The New York Times carried the story front page. Penzias and Wilson were awarded the Nobel Prize in 1978. In December of 1965 Dicke's assistants Wilkinson and Roll got their rooftop antenna working and were able to confirm the Bell lab's result.

COBE

Detecting the existence of the microwave radiation turned out to be just the start of a twenty-four year odyssey. Scientists wanted to know more about this soft glow that seemed to fill the whole universe. They needed to measure its temperature more accurately to see if there were any variations which would give more information about the nature of the Big Bang.

The main problem was water vapour in the Earth's atmosphere. Researchers went to the driest deserts they could find with their equipment, but even the desert air had too much water. They tried getting above the moisture by setting up on White Mountain, 12 500 feet (3 800 metres) above California. Others lifted their instruments using high altitude balloons which could reach three times higher than Mount Everest. They tried rockets used by meteorologists, and even high altitude U-2 spy planes. The experiments were only partially successful.

What was needed was a satellite that could take the equipment far above the atmosphere to an orbit 900 kilometres away from the Earth.

Just such a chance presented itself in the summer of 1974 in the shape of an 'Announcement of Opportunity' put out by NASA, the U.S. space administration. It was looking for suggestions for new space missions. John Mather, the son of a farmer in New Jersey, who worked for NASA, convened a meeting at the Goddard Space Science Centre in upper Manhattan, and proposed a satellite with instruments to provide an in-depth analysis of the background radiation.

Because it is funded by the U.S. government, NASA is quick to choose projects that will be popular with the public, and this one would show where everything in the universe came from. It would reveal the seeds out of which the galaxies had grown. Two other teams put in similar proposals, and out of the three came a satellite known as the Cosmic Background Explorer, or COBE for short. At first it was planned to put the satellite in orbit using a relatively inexpensive Delta rocket, but NASA insisted on using the Space Shuttle, even though it meant cancelling the best orbit which was high above the polar regions. NASA gave the go-ahead in 1982, setting a launch date for 1989. Then when COBE was nearing completion on 28th January 1986, the Challenger disaster occurred, which killed seven astronauts. The whole Space Shuttle programme was frozen.

John Mather was reluctant to accept defeat and persuaded NASA to use the much smaller Delta rocket. The trouble was this meant re-designing and re-building COBE. From being a 10 500 pound (4 760 kg) object, it had to be reduced in size and weight to 5 000 pounds (2 250 kg).

Finally, on 18th November 1989, COBE was launched from Vandenberg Air Force Base, not far from Los Angeles. Mather invited along Ralph Alpher and Robert Herman who had been the first to estimate the left over temperature of the Big Bang all those years ago. All went well, and COBE began to measure the temperature left over from the Big Bang.

What everyone wanted to know was whether the radiation conformed to what was known as a black body curve. One commentator said that as Mather put up his results on a screen there was complete silence. When it became clear that COBE had discovered the most perfect match anyone had ever seen, the audience began to buzz, then clapping broke out, and suddenly everyone was on their feet giving a resounding standing ovation! A very uncharacteristic response at a formal scientific meeting.

Marcus Chown says, "COBE had seen to the very heart of things. It had stripped away all the bewildering complexity of the Universe. And there at the beginning of time was breathtaking simplicity – more beautiful than anyone had dared imagine."[6] The temperature of the radiation was confirmed as 2.726 degrees above absolute zero.

Ripples from the Beginning

COBE's second triumph came two years later. One of the problems with the Big Bang theory was that it predicted a smooth universe in all directions, yet when astronomers mapped the whole sky for galaxies they found they were not evenly spread out. They took the shape of interconnected chains of galaxies with big gaps in between. There were great voids of empty space between the lace-like clusters, which were then often linked to super-clusters, and so on. For these structures to evolve there must have been 'micro lumps' or ripples in the very early universe which should have left their mark on the microwave background.

One of the COBE's detectors, known as a Differential Microwave Radiometer (DMR) was searching for these variations. On 23rd April 1992 at a meeting of the American Physical Society, George Smoot, who led the radiometer team, described to a shocked audience how COBE had managed to detect tiny temperature ripples in the background radiation.

There was a massive media response to this discovery. Newspapers all over the world ran headlines like, 'How the Universe was Made!'

Stephen Hawking was quoted as saying, "It is the scientific discovery of the century, if not of all time."[7] George Smoot said, "If you're a religious person, it's like seeing the face of God."[8]

Eight months later another experiment was carried out by a team of scientists from the Massachusetts Institute of Technology. They used a high altitude balloon at 25 miles (40 kms) with a newer detector that was twenty-five times more sensitive than COBE's. It confirmed the earlier results.

References

1. Hawking, Stephen. 1988. *A brief history of time.* London: Bantam Press. p 6.
2. Smoot, George., and Davidson, Keay. 1993. *Wrinkles in time.* London: ABACUS. p 54.
3. ibid., p 78.
4. Chown, Marcus. 1993. *Afterglow of creation.* Reading: Arrow Books. p 38.
5. ibid., p 64
6. ibid., p 125
7. Fraser, Gordon., Lillestol, Egil., and Sellevag, Inge. 1994. *The search for infinity.* London: Michell Beazley. p 137.
8. ibid., p 137.

Part Three

Exploring the Smallest

Chapter Ten
The Atom

The Disadvantage of Common Sense

If there is one thing that science has shown us, it is that using our ordinary common sense to discover the truth about the world is not a very good idea. I was born in Central Africa, and the landscape around where I grew up in Matabeleland was flat, except for a few granite outcrops and rolling hills. I can still remember at a young age the jolt I experienced when I was told that the Earth is actually round!

To me it was obviously flat. If you travelled far enough beyond the horizon you would almost certainly fall off the edge of the world. When it was explained to me, I remember feeling a sense of wonder. I can't help thinking that the Apostles would have felt the same. Later I could even sympathise with the members of the Catholic Inquisition who persecuted Galileo. The idea that the Earth moved round the Sun must have seemed quite ludicrous. Everyone witnessed with their own eyes a daily sunrise and sunset, and it was clearly the Sun that was moving across the sky!

Apart from anything else, as the Inquisitors pointed out, no matter how hard you tried, you could not feel the Earth moving beneath your feet. Imagine, then, the impact that Isaac Newton must have had on people's minds when he worked out how gravity explained the movement of the planets and how the ocean tides are created by the moon. No wonder he was a huge celebrity. It must have seemed like a mist of darkness had suddenly lifted. The story of scientific discovery seems to be continually showing us how amazingly intelligent Nature is. Each time we fumble forward to a new discovery, a brand new horizon opens up to amaze us.

Well, the world of the atom presents us with what are the most shocking abuses of common sense anywhere in science. For a start, they are extremely tiny. One hundred million placed side by side would stretch across a single centimetre on your old school ruler. But what is 100 million? For that matter, how big is one million?

When full, a really big sports stadium can hold nearly 100 thousand people. So, if you imagine ten sports stadia lined up next to each other, then that would be one million people. You would need to increase this number by a hundred, in other words, one thousand sports arenas, to make up 100 million people. That's a lot of bits to fit along the first centimetre on your old school ruler!

If you think of them in a lump, not next to each other in a row, then your average grain of fine table salt contains roughly fifty million atoms. Now, think about the biggest open air rock festivals like Glastonbury that attract perhaps half a million people; then, if each person was an atom, it would take a hundred big festivals to make up one grain of table salt. A lot of objects for such a tiny space.

Although so tiny, most people have an idea of what an atom looks like. We imagine it like a very miniature solar system, with the Sun representing the nucleus or centre of the atom, which is a tight bundle of protons and neutrons, and the electrons orbiting the nucleus like planets around the Sun.

The main difference being that the electrons can jump from one orbit to another. And another difference is that they are zipping around the nucleus at a screaming rate so that they would look more like a cloud.

But the biggest surprise, is the discovery that although an atom is so small, it is in fact mostly empty space! Going back to the comparison with sports arenas, if the nucleus of the atom were a grain of rice floating in the middle of the pitch, then the electrons, which are much smaller and lighter, would be buzzing around it at the outer edges of the furthest stands. Another example that is sometimes given goes like this. If the nucleus was the size of a golf ball then the closest electron orbit would be roughly one kilometre from the golf ball!

THE ATOM

It is quite shocking to realise that everything so familiar to us, from our fingers and hands, and our household furniture, as well as the buildings we live in, are all made up out of atoms, yet they are themselves largely empty space. Just when you are getting to grips with what you think is the final tiniest bit of matter – which you expect to be solid – you discover that it is empty on a massive scale.

Sizes

Talking about scale, what numbers are used to measure things down to the size of an atom? The nail on your pinkie finger is roughly one centimetre across. If we stick to centimetres, then the thickness of your skin is about a tenth of a centimetre, which can be written 10^{-1}. A cell in your skin is about 10^{-3}cm, a virus about 10^{-5}cm, a DNA molecule 10^{-6}cm, and an atom 10^{-8}cm.

1 centimetre	1	pinkie nail
10^{-1}cm = one tenth cm	$1/10$	thickness of skin
10^{-2}cm = one hundredth cm	$1/100$	
10^{-3}cm = one thousandth cm	$1/1\,000$	human cell
10^{-4}cm = one ten thousandth cm	$1/10\,000$	
10^{-5}cm = one hundred thousandth cm	$1/100\,000$	a virus
10^{-6}cm = one millionth cm	$1/1\,000\,000$	DNA molecule
10^{-7}cm = one ten millionth cm	$1/10\,000\,000$	
10^{-8}cm = one hundred millionth cm	$1/100\,000\,000$	an atom

Similarly, we can go the other way – upwards. If we take the length of our index finger as being about 10 centimetres long, then:

10^1cm is 10 x 1 = 10 : ten cm: index finger
10^2 cm is 10 x 10 = 100 : a hundred cms or 1 metre
10^3cm is 10 x 10 x 10 = 1 000 : one thousand cm : 10 metres
10^4cm is 10 x 10 x 10 x 10 = 10 000 : ten thousand cm = 100 metres : football pitch
10^5cm is 10 x 10 x 10 x 10 x 10 = 100 000 : one hundred thousand cm : 1 kilometre

and so on. If we keep sticking with centimetres, then the distance to the sun is 10^{11}cms, the size of our galaxy 10^{23}cm, and the limits of the universe at 10^{29}cm.

Although convenient, powers of ten can be very misleading. They double in size each time, and each time they double, they then double again, what maths calls exponential. Like 2^2 is 2x2=4 but 4^2 is not 4x2=8, it's 4x4=16. Jean Heidmann gives this good example: "Suppose a court of law handed down a sentence of 10^8 seconds in prison and the defence appealed for a reduction of 10^7 seconds; perhaps the court would not trouble to discuss for very long such a trifle, since the difference between 7 and 8 is not much. But 10^7 seconds is four months whereas 10^8 seconds is three years, quite an enormous difference for the prisoner!"[1]

Layers of Reality

As if the size of an atom weren't small enough, the layers of reality actually go on and on to much deeper levels. Inside the protons and neutrons within the grain of rice that is floating in the middle of the stadium, there live objects called quarks. On an even deeper level than quarks you reach the strange land of 'super strings'. They exist in a bizarrely different world from ours where the norm is ten or eleven dimensions. Sizes at this level are measured by what is called the Planck Scale. Stephen Weinberg, our main guide, says, "Planck's constant is 6.626 thousandth millionth millionth millionth millionths, a decimal point followed by twenty-six zeros and then 6626."[2] The Planck time is 10^{-44} of a second which is sometimes rather comically referred to as a 'jiffy'. It is the shortest 'blip' of time we know of so far. The Planck length is 10^{-35} of a metre. To get some idea of how tiny this scale of things is, the nucleus of an atom – just the grain of rice in the middle of the stadium, is as big to a super string as the whole universe is to us. That means it's pretty tiny! It really makes you wonder if the layers of reality might go on and on forever.

Instead of point-like bits, super strings are like tiny loops of vibrating 'string'. Just like a violin string they can vibrate at different frequencies and each different frequency is equal to a particular kind of subatomic particle like a quark or an electron. Freeman Dyson, who is Professor of Physics at the Institute for Advanced Study in Princeton (the post held by Einstein most of his working life), describes the size of superstrings like this:

> They are small. They are extravagantly small....Imagine if you can, four things that have very different sizes. First the entire visible universe, second, the planet Earth. Third, the nucleus of an atom. Fourth, a superstring. The step in size from each of these things to the next is roughly the same. The Earth is smaller than the visible universe by about twenty powers of ten. An atomic nucleus is smaller than the Earth by twenty powers of ten. And a superstring is smaller than a nucleus by twenty powers of ten.[3]

So a super string sees the nucleus of an atom (our grain of rice), like we see our entire visible universe. All 13.7 billion light years of it!

How Far can we Go?

This level is so extraordinarily deep for the human mind to comprehend that it makes you wonder if we are not looking at something which is infinite. Does our reality go on and on forever getting smaller and smaller? It begs the question, at what really profound level do things stop. If they ever stop. It's an adventure we will be exploring later.

If you want a shiver down your spine, consider the possibility that we are somehow creating reality as we explore it? Does our consciousness somehow alter how the universe is constructed?

As far as super strings are concerned, they are strictly speaking still at the purely theoretical stage. To test for the existence of super strings would require technology well beyond what exists at the moment. The reason people have been so excited about super string theory, also known as M-Theory, is that it seems to offer the first ever possibility of finding an explanation of the universe.

THE FINAL MYSTERY

For most of the last century our ideas about how the universe is constructed have been dominated by two major theories. The first one is Einstein's Theory of General Relativity which explains the very large part, from planets and stars to galaxies and the whole universe, and the second is Quantum Theory, which explains everything on the smallest scale.

If it were possible to link these two theories together then we would have a Theory of Everything. Super string theory looked like being a promising candidate. In his Dirac Memorial lecture Steven Weinberg said, "This theory has the smell of inevitability about it."[4]

The trouble is super strings are on such a deep level that they will never be seen. Another small disadvantage is that they inhabit a ten or eleven dimensional space, which is a bit of a nightmare to try and visualise. I should say though that hundreds, even an infinite number, of dimensions are apparently a piece of cake for mathematics. Is that telling us something?

A similar thing happens when we start looking into time. What, for example, is the present moment? How long is 'now'? Does it last one second, or half a second, or a tenth of a second? Or the smallest moment in time for physics as mentioned above, the Planck time. Paul Davies a Professor of Astrophysics at the University of Arizona says, "there are no less than one followed by forty three zeros (written 10^{43}) of them in one second, a duration so short that even light can travel a mere million-billion-billionth of a centimetre in one jiffy."[5] And light is the fastest thing in the universe.

Or perhaps both time and space can be divided infinitely? Perhaps we are surrounded by infinities and don't know it? After all, we aren't equipped to see radio waves either. Freeman Dyson once wrote a book called 'Infinite in All Directions'. Could it be that the infinities define how it is that things become real? It's a bit scary, but it's also exciting.

References

1. Heidmann, Jean., 1989. *Cosmic odyssey.* Cambridge: Cambridge University Press. p 20.
2. Weinberg, Steven., 1993. *Dreams of a final theory.* London: Vintage. p 57.
3. Dyson, Freeman., 1990. *Infinite in all directions.* London: Penguin. p 18.
4. Weinberg, Steven., 1989. *Elementary particles and the laws of physics.* Cambridge: Cambridge University Press. p 104.
5. Davies, Paul., 1990. *Other worlds.* London: Penguin. p 95.

Chapter Eleven
Crossing the Atomic Boundary

Starters

We need to go back a bit to find out how we got to know about atoms. It's an interesting journey because it involves many people. The idea of atoms was first put forward by a classical Greek philosopher (yes, the Greeks again!) called Democritus, who lived in Abdera about 450 BC. His proposal was that the universe contained only a vacuum of nothing, and atoms. His atoms were invisibly small and hard, and constantly in motion, amazingly close to the modern picture. He saw atoms also as 'uncuttable' and indestructible; they were the smallest units out of which everything was built up.

He travelled a lot in his life, visiting Egypt and Persia. The story goes that some of his neighbours thought he was mad, so they sent for Hippocrates, the famous doctor (responsible for the Hippocratic Oath). After some time Hippocrates returned from the visit, saying he had never met anyone more sane!

Plato and Aristotle did not favour Democritus's ideas, so he never became part of the mainstream of Greek philosophy. They preferred the more primitive idea that everything is made up of four primary substances: earth, air, water and fire.

Because the Catholic Church adopted Aristotle's model, Democritus dropped out of sight. His work was probably known to Isaac Newton and Robert Boyle, but modern atomic theory is generally considered to begin with John Dalton.

Discovering the Elements

An element is a substance like hydrogen or oxygen or carbon or iron. There are 92 of them that occur in Nature. They cannot be changed into another substance because each of them is made up of exactly the same kind of atoms. People didn't know this of course to begin with. John Dalton, who you might have heard of at school, was the first to come up with the idea. He was born in the English Lake District but worked in Manchester.

He suggested through his experiments that each different element had its own kind of atom, which was identical and indivisible. He made an attempt to measure the atomic weights of different elements, which he published in 1807. Although it was not accurate, it identified twenty different elements, from hydrogen, which is the lightest and simplest type of atom, to mercury. Uranium is actually the heaviest but Dalton didn't know about it. This is why the military use depleted uranium in artillery shells.

Chemists continued experimenting, and slowly more and more elements were discovered. It began to look as though a pattern might be happening in the different atomic weights. It was a Russian with the wonderful name of Dimitri Ivanovich Mendeleev, who came up with what is known as The Periodic Table in 1869. It was a way of classifying all the elements according to their atomic weights. He found that elements with similar properties occurred at regular intervals or 'periods', and that they could be set out in a table of vertical columns. In order for the table to work, he had to leave gaps in it. This turned out to predict the existence of elements which had not yet been discovered. When these elements *were* discovered over the years, it was a major triumph for Mendeleev's table and Dalton's atomic theory.

Like so many of the characters who pushed the boundaries of science, Mendeleev had an interesting life. Born in Siberia, he lived most of his life in St. Petersburg. He became Professor of Chemistry there, and gained a world-wide reputation. Although he was married, he fell passionately in love with a younger woman and begged her to marry him, which she did. The only problem was that he was already married. So he spent some years as a bigamist. Like Newton he did a lot of work for the government, particularly in standardising weights and measures. As a famous chemist he was asked to standardise how Vodka was to be produced throughout Russia. He decided, probably by trying many a tipple, that the best mix would be 40% alcohol to 60% water, so next time you knock back an iced Vodka, think of old Mendeleev.

Electrons

Because electrons are the bits of atoms that whizz about the nucleus of the atom, we need to know how they were first discovered. Although they knew that each element was made up of the same kind of atom, they had no idea what an atom looked like, or even how big it was. The story of electrons begins around 600 BC, long before Democritus. Another Greek philosopher, Thales of Miletus, whom we met before, discovered that strange lumps of iron ore, from a place called Magnesia in Asia Minor, were able to stick together by a force of attraction and were difficult to pull apart. He also discovered that a piece of amber or resin, when rubbed against his clothing, was able to attract feathers and light objects, and make his hair stand on end; what we now call static electricity. The word for amber in Greek was 'elektron'.

Then, taking a bit of a jump forward, in 1752 Benjamin Franklin in the U.S. was able to show that lightning is electricity, but everyone thought that magnetism and electricity were entirely different kinds of forces. You would hardly think that forked lightning and fridge magnets had anything in common. Their appearance is so completely different, so different that you would have been considered weird if you had said they were connected. Apart from gravity, which was obvious, these were the only other forces thought to exist in Nature, and they seemed distinctly separate.

Then in 1819, just after the Napoleonic wars, a Danish physicist, Hans Kristian Oersted, discovered to everyone's surprise that an electric current could make a magnetic compass needle swing violently. The news spread and it attracted the attention of Michael Faraday, in

England. He was something of a 'natural'; a self-taught scientist. He came from humble origins, had only an elementary school education, and was then apprenticed as a bookbinder. Although he had hardly any training in mathematics, he was brilliant at visualising problems, and ended up being the first and perhaps only member of the Royal Society who never went to university.

He was fascinated by the experiment we all do at school where a piece of card is placed over a magnet and iron filings are sprinkled on top. When the paper is tapped gently, the lines of magnetic force around the magnet suddenly become clearly visible. In 1831 he discovered how a moving magnet could produce an electric current, and went on to build the first electric dynamo.

Today his dynamo is used on a grand scale to create the electricity that powers our modern cities. When Prime Minister Gladstone asked Faraday of what use he thought electricity would be, Faraday is said to have replied that he did not know, "but he was sure the Government would one day figure out how to tax it."[1]

A Triumph for Maths

The next character in the story was James Clerk Maxwell, born in Edinburgh in 1831. He was something of a prodigy and had an invention published before he even left school. He became a Professor at Aberdeen University in Scotland at the age of twenty-six and was the winner of many academic prizes. He was the first person to predict that Saturn's rings were made up of small pieces of matter orbiting the planet. He later became a Professor at King's College in London, before retiring at the age of 39 to Glenair outside Edinburgh. But he came out of retirement when he was offered the first Professorship at the now world famous Cavendish Laboratory at Cambridge University.

What Maxwell did was to take the experimental facts about electricity and magnetism that had been discovered by Faraday and others, and try to describe them mathematically. When he did this the equations told him something extraordinary. Electricity and magnetism were part of the same force! He called the new force 'electromagnetism'. The magnitude of his discovery is difficult to fully appreciate today, but what he had done was to unite two out of the three forces which were then known in the universe. Physics has never been the same since. Most of the last century was spent trying to unite the other forces of Nature.

Whilst this was an enormous achievement, there was an even greater surprise waiting offstage which nobody even in their wildest dreams had thought possible. When Maxwell tried, just out of personal curiosity, to discover the speed at which his new electromagnetism travelled, he found that it was equal to 3×10^8 metres per second. But from optical experiments this had already been shown to be the speed of light! He must have taken a deep breath before he wrote:

> We can hardly avoid the inference that light consists in the transverse undulations of the same medium which is the cause of electrical and magnetic phenomena.[2]

Maxwell had managed not only to unite electricity and magnetism, but also to show that light was part of the same thing. His electromagnetic theory, first published in 1867, was a huge triumph, equal to Newton's theory of gravity. Steven Weinberg says, "Maxwell is generally regarded as the greatest physicist between Newton and Einstein."[3]

Just like Newton's law of the inverse square before, Maxwell's five equations are a major confirmation of the frightening power of mathematics to describe our physical world. The equations revealed something about the physical world which no-one including Maxwell had ever suspected, namely that electricity and magnetism were different aspects of a single phenomenon; but not only this, that electricity and magnetism are the same thing as light.

Maxwell went on to predict that electromagnetic waves come in many different wavelengths. What the human eye is able to see is only a tiny part of the whole electromagnetic spectrum. Evolution has equipped us to see only wavelengths which are less than a thousandth of a millimetre, which we call light. The shortest wavelengths of all are known as gamma rays, and the longest as radio waves. As we saw in part one, the speed of this stuff is a shocking 300 000 kilometres or 186 000 miles per second.

In 1887 Heinrich Hertz, the Professor of Physics at Karlsruhe in Germany, produced the first radio waves, thereby verifying Maxwell's theory, and proving that we live in a universe which is filled with electromagnetic radiation, almost all of which is invisible to the human eye. It includes radio waves, television, radar, microwaves, infra-red, ultra violet, x-rays and gamma rays.

Getting inside the Atom

In 1891, George Stoney, an Irish physicist working in Dublin, had suggested there might be a 'smallest unit' of electricity, which he called the electron. But it was at the famous Cavendish Laboratories in Cambridge that the atomic boundary was finally crossed.

It was Joseph John Thomson, known to everyone as J.J., who actually discovered the electron. He was the son of a bookseller and had wanted to be an engineer but could not afford the money that was then needed to be an apprentice. He ended up studying maths, physics and chemistry instead, and won a scholarship to Cambridge in 1876, later becoming a Fellow of that famous temple of learning, Trinity College.

In 1884 he was elected to the Cavendish Professorship of Experimental Physics. Ironically, he is well known to have been bad at actually doing experiments. One of his assistants remembered that, "J.J. was very awkward with his fingers, and I found it necessary not to encourage him to handle the instruments."[4]

These were 'state of the art' at the time. His team managed to create a vacuum inside a glass tube that could reduce the pressure to one ten-thousandth of normal atmospheric pressure; the best achieved up until that time. What he did was to fire a beam of electricity through this vacuum which created what were called Cathode rays. By deflecting the path of the beam using magnets outside, he was able to measure both the size and the amount of charge of each 'corpuscle', as he first called them. He proved that these electrons were negatively charged and about two thousand times smaller than an atom of hydrogen.

Thompson realised that he had discovered things that were bits of atoms, and that they had never been seen before.

> We have in cathode rays matter in a new state, a state in which the subdivision of matter is carried very much further than in the ordinary gaseous state: a state in which all matter – that is matter derived from different sources such as hydrogen, oxygen, etc., – is of one and the same kind; this matter being the substance from which the chemical elements are built up.[5]

With the discovery of the electron, the theory of Democritus, that atoms were indivisible, came to an end. Physics had crossed the barrier of the atom and was now inside the world of subatomic particles.

J.J.'s work was first published in 1897 and he was awarded the Nobel Prize in 1906 for the discovery of the electron. In 1908 he obtained a knighthood and became President of the Royal Society in 1915. He was appointed Master of Trinity College in 1918, and when he died he was buried in Westminster Abbey not far from Sir Isaac Newton.

During his leadership of the Cavendish Laboratories he managed to attract students of the highest quality from all over the world. Amongst them was a New Zealander, Ernest Rutherford, whom we turn to next.

References

1. Lederman, Leon., and Schramm, David. 1989. *From quarks to the cosmos.* New York: Scientific American Library. p 231.
2. ibid., p 25.
3. Weinberg, Steven., 1993. *The discovery of subatomic particles.* Harmondsworth: Penguin. p 7.
4. ibid., p 13.
5. ibid., p 66.

Chapter Twelve
The Quantum Revolution

As Luck Would Have It

Rutherford was born in Brightwater in New Zealand and, after studying at Canterbury College in Christchurch, he won a scholarship to Trinity College. He arrived in 1895 at the age of twenty-four in time to witness the discovery of the electron.

He first worked on radio waves, and for a time held the world record for transmission over a distance of three kilometres before he was beaten by Marconi. But he soon changed over to working on something new called 'radioactivity', which had just been discovered and quite by accident. Another of those lucky accidents in science.

Antoine Henri Becquerel, Professor of Physics at the Ecole Polytechnique in Paris, had been doing research on different minerals and how they reacted in sunlight. It just so happened that the weather in February was wetter than usual. So, a bit fed up because there was too little sun, he wrapped up his materials which consisted of a photographic plate and some uranium salts, in some black paper and put them away in a chest of drawers.

On Sunday 1st March the sun began to shine again, and before continuing with the experiment he decided on a whim, to develop the photographic plate. To his surprise, he found that despite being in total darkness the plate was covered with traces of invisible rays. He guessed that the uranium salts were to blame and called them uranium rays.

Rutherford decided to study these rays to find out what they were. He discovered that they were made up of two different types, which he called alpha and beta. They turned out not to be rays of the electromagnetic spectrum which is what most people expected, but actually tiny high-speed particles ejected by uranium in the natural process of what became known as radioactive decay. It was a great discovery, something quite new to science.

Shortly after this, Rutherford was offered the position of Research Professor at the brand new Macdonald Physics Laboratory at McGill University in Montreal, Canada. He set sail with a few bags of uranium and thorium salts (another radioactive substance) in his pocket, so to speak, so he could continue doing his experiments. In those days no one even thought about the dangers of radioactivity like we do today. In fact it is amusing to think that had it been the 21st century he would almost certainly have been arrested as a terrorist suspect.

At McGill he carried out experiments to discover exactly what these high speed particles were. The alpha particles turned out to be ions of the gas helium - that is, atoms of helium with their electrons stripped away; and the beta particles turned out to be Thomson's electrons.

In 1906 Rutherford's success led to him being offered a Professorship at Manchester University where Faraday had worked. Since he felt isolated from developments in Europe, he decided to return to Britain.

Experiments in the Dark

Because atoms were far too tiny to be seen, people had no idea what they looked like. All they knew was that electrons were even smaller bits of the atom. Rutherford knew that his alpha particles were ejected from radioactive materials at high speed, around 10 000 kilometres per second, so he decided to use these as bullets to shoot at atoms to see what happened.

He used a thinly beaten piece of gold foil as the target atoms, and surrounded this with screens which were painted with zinc sulphide. When a particle hit the zinc sulphide it would release a series of 'photons' or particles of light. These flashes could only just be seen with the human eye. Rutherford and his students had to sit in total darkness for several hours to adapt their eyes so that they could see these tiny flashes and record them. They also had to repeat the experiments many times to give reliable results.

They were shocked by what happened. As expected, most of the alpha particles passed straight through the gold foil and produced a flash on the screen behind. But every now and then, about one in 8 000 of the particles would bounce straight back as though it had hit something tremendously hard. Rutherford was stunned: "...quite the most incredible event that has ever happened to me in my life...It was almost as if you fired a 15 inch shell at a piece of tissue paper and it came back and hit you."[1]

What he had done was to discover the nucleus of the atom. The consequences of this experiment, made in a darkened room in a Manchester laboratory in 1911, were to be very far reaching as you can imagine.

From the experiments, Rutherford and his team were able to work out that the atom consisted mostly of empty space as we have seen, but deep in the centre was the positively charged nucleus, which contained most of the mass of the atom.

Discovering the Proton

In 1919 he moved from Manchester to succeed Thompson at the Cavendish as Professor of Experimental Physics. It was here that he sent his alpha particles zipping into nitrogen gas and found that the collisions knocked a positively charged particle out of the nitrogen atoms. He guessed that this was a fundamental bit of the nucleus and called it the proton or 'first particle'.

This achievement was of great importance. Not only had he discovered the proton, but in knocking a proton out of a nitrogen atom he had converted it into an atom of oxygen. Changing one element into another had been the dream of the ancient art of alchemy which Newton and all those before him had taken so much time and effort to master.

Rutherford and his team were the first humans to witness this fundamental transformation created artificially in the laboratory.

During the First World War, Rutherford served on a submarine committee, and when he missed one of its meetings he gave this legendary excuse: "If, as I have reason to believe, I have disintegrated the nucleus of the atom, this is of greater significance than the war."[2] He knew his priorities!

He should have won several Nobel prizes, but in fact he won only one, not for physics but for chemistry, in 1908. It was for his work on radioactivity and identifying alpha and beta particles. This was ironic, because he was contemptuous of the subject. He once said that all science is either physics or stamp collecting. He was a big man with a booming voice that could actually upset delicate instruments. Because of this, his students had a large sign hung in the middle of the laboratory, which read: 'Talk Softly Please'. Despite his remarkable success he made some wonderful clangers. He once dismissed the possible use of nuclear energy as "moonshine".

Apart from his own work, many of his students became famous. In 1925 Patrick Blackett invented the cloud chamber, which made it possible to photograph the tracks of particles in subatomic collisions, so that there was an actual record of these events. He won a Nobel Prize in 1948 for this work.

John Cockroft and Ernest Walton invented the first ever particle accelerator, which was the forerunner of the giant machines of today. They shared the Nobel Prize in 1951, and in 1932 James Chadwick, Rutherford's principal researcher, discovered the neutron, which had been predicted by theorists for some time. He received a Nobel Prize in 1935.

Rutherford was knighted in 1914, became President of the Royal Society in 1925, and was made a Peer of the Realm in 1930, becoming Lord Rutherford of Nelson. For part of his coat of arms he chose kiwi birds, showing that he had not forgotten his roots in New Zealand.

The New Age

Brilliant as Rutherford's discoveries were, his model of the atom had two serious flaws. Thomson had shown that electrons were negatively charged particles; and Rutherford, that protons in the nucleus were positive. Just like two ends of a magnet, they should have attracted each other, so that the electrons, being much lighter, should be immediately sucked into the nucleus. Secondly, according to Maxwell's electromagnetism, the moving electrons should give off radiation, thereby losing energy and quickly falling into the nucleus. Yet atoms are manifestly stable; the whole of the universe is constructed out of them. If they collapsed we couldn't exist.

This problem marks a turning point in the history of science. Everything up to this point is now known as classical physics; everything after is called modern physics. The revolution that took place at this turning point has provided us with all the familiar technology of our everyday use, from lasers and iPhones to mobiles, CD's, HD television, satellite navigation and so on. It marks the transition between the industrial revolution and the technological revolution, and made possible the information age in which we live. These changes would certainly never have happened if it were not for the Quantum Revolution. Because it dwarfs into insignificance other events like the French Revolution, I believe that it fully justifies the use of capital letters. It is undoubtedly the biggest revolution ever to happen so far.

Its origins can be conveniently dated to the year 1900. On the 14th December of that year, Max Planck presented a paper to the German Physical Society which showed that heat radiated from a hot object such as a red hot poker, was given off in tiny packets of energy which he called 'quanta', from the Latin meaning 'how much'. This is the origin of the Planck Scale that we were talking about earlier. Up until then, scientists had been convinced that heat radiation came in waves. How could it also come in bits?

An Act of Despair

Planck had been trying to solve a problem that many of the best minds of the time had tried to solve, but without success. It was only when, in desperation, Planck tried a different mathematical approach, pioneered earlier by a man called Ludwig Boltzmann, that the problems suddenly disappeared. He said, "I can characterise the whole procedure as an act of despair…a theoretical interpretation *had* to be found at any price, however high it might be."[3]

The high price that Planck referred to was the passing away of classical physics. He was a deeply conservative man and did not like what he had discovered. He had been as convinced as everyone else that heat radiation was a wave. But the problem could only be solved by assuming that the waves were somehow also packets or particles of energy.

Once again this is yet another iconic example of the extraordinary power of mathematics to tell us absolutely fundamental things about the world. In this case it was again something utterly profound that absolutely nobody wanted to know or had ever suspected. Although it did not cause much of a stir at the time, Planck realised the enormity of what he had done. Werner Heisenberg, the discoverer of the famous Uncertainty Principle, was told a story by Planck's son that after an intense period of work in the summer of 1900, his father spoke to him during a long walk that they took together in the Grunewald, a wooded park in Berlin. "On this walk he explained that he felt he had possibly made a discovery of the first rank, comparable perhaps only to the discoveries of Newton."[4] It doesn't seem much at first but wait and see what's coming!

Planck's personal life was filled with tragedy. His eldest son was killed at the Battle of Verdun in the First World War, two of his daughters died in childbirth soon after marrying, and his youngest son Erwin was executed by the Nazis for his part in the attempted coup against Hitler in 1944. Einstein said of Planck, whom he held in the highest regard: "his work has given one of the most powerful of all impulses to the progress of science. His ideas will be effective as long as physical science lasts."[5]

The Revolution Begins

Because of its huge implications, the problem that Planck had discovered was at first thought to be just a mathematical trick. However, five years on, in 1905 Albert Einstein, working in the Swiss Patent Office in Bern, published three famous papers in one and the same edition of the scientific journal 'Annalen der Physik'. As we mentioned in part one, the first was about a phenomenon called the photoelectric effect, which proved beyond doubt that light was indeed made up of particles rather than waves. Later called photons. Plank's quanta were real. Einstein won the Nobel Prize for this discovery in 1921.

With his work on the photoelectric effect, Einstein proved that Planck's quanta were not just a convenient mathematical trick, but actually real. Light was both a wave and a particle! This is not just the case for photons of light, all particles of matter are both waves and particles at the same time, as we will see in the next chapter. It sounds very strange, but this is the signature of the quantum world.

The greatest revolution of all time had just begun.

References

1. Weinberg, Steven., 1993. *The discovery of subatomic particles.* Harmondsworth: Penguin. p 122.
2. Fraser, G., Lillestol, E., and Sellervag, I. 1994. *The search for infinity.* London: Michel Beazley. p 38.
3. Gribbin, John., 1984. *In search of Schrödinger's cat.* London: Corgi Books. p 42.
4. Heisenberg, Werne.r, 1959. *Physics and philosophy.* London: Allen & Unwin. p 35.
5. Fraser, G., et. al., op. cit. p 30.

Chapter Thirteen
The Age of Uncertainty

Saving the Atom

It was Planck's quantum of energy that came to the rescue of Rutherford's model of the atom with its electrons that should have been sucked into the nucleus. It arrived in the shape of a young Danish physicist from Copenhagen named Niels Bohr who was to become one of the most famous scientists of the twentieth century. He and his brother Harold played football to professional level in their youth. Niels kept goal and his brother, who was a mathematician, played halfback. That must be some kind of record.

After completing his doctorate in 1911, Niels went to England to join J.J. Thomson at the Cavendish where he stayed for eight months. He and J.J. didn't get on too well, so he moved to Manchester to work under Rutherford for the next four years.

By applying Planck's formulae to the electron, he showed that it could occupy certain fixed orbits around the nucleus without it losing energy and falling into the centre. Radiation could only take place when the electron jumped from one specific orbit to another. When it did this it gave off a photon of light and jumped to a lower orbit, or absorbed a photon and jumped up an orbit.

Everything is One

In discovering the mechanism that kept atoms stable, Bohr had also explained what spectral lines were. At school you do an experiment where light from a hot gas is shone through a prism, and it comes out as a row of coloured bands, a bit like a supermarket bar code, except in different colours. Each gas has a different sequence of colours, so these lines provide a signature or bar code of whatever element it is that is being burnt. This is still the method astronomers use to tell the composition of stars that are burning millions of light years away.

I should pause here to point out that something big is beginning to happen here in our journey. When we last saw Einstein in Part one, we were talking about the huge cosmic world of the universe. Now, as we have seen, he also made a large contribution to the microscopic world of atoms. In discovering that spectral lines were made by the particular structure of each type of atom, Bohr was linking the tiny world of atoms with the massive world of stars and galaxies. As our story goes on, this link between the two worlds gets stronger and stronger, showing us that everything in the universe, from the biggest to the smallest, is all joined up together as an interdependent whole. Perhaps the biggest of the broad achievements of twentieth century science was this demonstration that everything which exists is one composite entity.

To continue, in 1916 Bohr returned to Copenhagen and in 1918 became the first director of its new Institute for Theoretical Physics, which quickly became a world centre for the subject. As quantum theory developed, Bohr became one of its leading gurus. He and Einstein carried out a long lasting argument over many years about the nature of quantum events, which we will look at more closely later. He won his Nobel Prize for Physics in 1922. Bohr's ideas influenced the majority of physicists for two generations, and the line of thinking he developed became known as the Copenhagen Interpretation of quantum physics. Many still cling desperately to this way of thinking because the only alternatives are so utterly bizarre that they reduce human comprehension to a soggy pulp. The exciting thing though, is that this is the direction where the truth now certainly lies; much more about this later.

Niels' mother was Jewish, and Denmark was occupied by the Nazis during the war. In the autumn of 1943 he was forced to escape in a fishing boat to neutral Sweden and was then flown secretly to England in the bomb-bay of a Mosquito aircraft. During the war he was responsible, along with other prominent physicists, for lobbying President Roosevelt and Prime Minister Churchill about the danger of atomic weapons.

The Great Mystery of Waves and Particles

The central problem still remained. As Heisenberg describes it, "How could it be that the same radiation that produces interference patterns, and therefore must consist of waves, also produces the photoelectric effect, and therefore must consist of moving particles?"[1] He goes on to say, "the strangest experience of those years was that the paradoxes of quantum theory did not disappear during this process of clarification; on the contrary, they became even more marked and more exciting."[2]

Although Bohr's atom worked fine for the simplest atom of one electron and one proton (hydrogen), it could not cope with anything more complicated, nor was it able to explain how the electrons actually jumped from one orbit to another. Heisenberg described it as, "a peculiar mixture of mumbo-jumbo and empirical success."[3]

Bohr knew his model was far from complete and invited Heisenberg to Copenhagen where they worked together. Then, in the summer of 1925, Heisenberg was returning home to Gottingen when he suffered an attack of hay fever and decided instead to go to the island of Heligoland to escape the pollen. There he set about trying to describe mathematically all the known facts about the behaviour of atoms. He drew up a table or matrix for each known quantity, and then interwove them, and found that it worked in describing atoms.

Amazingly, when he returned to Gottingen, he was told by a friend that matrix multiplication had been discovered much earlier by mathematicians who had no idea that it would ever be used to describe anything in Nature. About this, Steven Weinberg says, "this is one example of the spooky ability of mathematics to anticipate structures that are relevant to the real world."[4] It provides yet another major triumph for the daunting power of mathematics to describe the world around us.

Matter Waves?

A year before Heisenberg's discovery, a French aristocrat, Prince Louis-Victor de Broglie, had used his doctoral thesis to put forward the rather startling idea that if light could be both a wave and a particle, then perhaps ordinary matter like electrons and protons could also be waves.

De Broglie had first studied history at the Sorbonne in Paris, but became interested in physics while serving as a soldier in World War 1 when he had operated a radio station on the Eiffel Tower. He was awarded a Nobel Prize in 1929 after his idea was proved to be true.

I must apologise for repeatedly reminding you, but it is very important to where we are going. This is indeed yet another example of the power of mathematics. De Broglie's equations in his thesis at first seemed crazy to everyone; something that no one dreamed possible. I mean, how could a physical object like a proton possibly be a wave? Yet later experiments actually proved that the mathematics had been correct. You may well be thinking by now: what *is* maths then? John Barrow, the Professor of Mathematical Sciences at the University of Cambridge, has called it: 'the dark mystery at the heart of science'. It is a major theme that we will be exploring as we make our way along this journey towards enlightenment.

In 1927, J.J's son, George Thomson, working in Britain, and Clinton Davisson in the U.S., performed a beautiful series of experiments using crystals of nickel, which proved conclusively that electrons also behave as waves. Max Jammer, a science historian, was moved to write, "one may feel inclined to say that Thomson, the father, was awarded the Nobel Prize for having shown that the electron is a particle, and Thomson, the son, for having shown that the electron is a wave."[5] That has to be some sort of major confirmation that the electron (and, incidentally, all other matter) is both a wave and a particle. Thomson and Davisson shared the Nobel Prize in 1937.

Over Christmas of the same year that Heisenberg published his paper on the quantum matrix (1925), an Austrian physicist, who was interested in de Broglie's 'matter waves', decided to take a skiing holiday in the Swiss Alps with his girlfriend, leaving his wife behind in Zurich. However, he never got much skiing done, not because of his girlfriend, but because he was 'distracted by a few calculations'.

These calculations turned out to be a new and alternative method of describing the atom. It was called wave mechanics and its founder was Erwin Schrödinger. In his model of the atom, the electrons were confined within standing wave patterns surrounding the nucleus. Like vibrations on a violin string, only complete wave patterns could fit, which explained the quantum jumping in Bohr's model.

Things were really getting to be very strange. Weinberg gives us a glimpse of this strangeness when he says, "neither de Broglie nor Schrödinger nor anyone else at first knew what sort of physical quantity was oscillating in an electron wave."[6] He reinforces this later by emphasising that no one knew what actual physical phenomenon "was being described by these numbers."[7]

Then, shortly after Schrödinger published his famous equation, Max Born, who was Heisenberg's mentor and professor at Gottingen, first proposed treating the waves in terms of probability. Probability? Weinberg goes on to explain, "in other words, electron waves are not waves of anything; their significance is simply that the value of the wave function at any point tells us the probability that the electron is at or near that point."[8] In a similar vein Heisenberg says, "This concept of the probability wave was something entirely new in theoretical physics since Newton."[9] Something more like a mathematical wave than a real one? Things were beginning to get a bit spooky.

The Uncertainty Principle

Our view of reality was in turmoil. Heisenberg recalls, "I remember discussions with Bohr which went through many hours till very late at night and ended almost in despair and, when at the end of the discussion I went alone for a walk in the neighbouring park, I repeated to myself again and again the question: can nature possibly be as absurd as it seemed to us in these atomic experiments?"[10] Worse was in store for Heisenberg when, the following year (1927), he actually confirmed the absurdity by means of his now famous 'Uncertainty Principle'. In struggling to form a mathematical description of an electron's path through a cloud chamber, he discovered what has become one of the most dramatic truths about the physical world ever discovered.

The 'Uncertainty Principle' states that it is impossible to know exactly at the same instant both where a particle is and how fast it is moving. Stephen Hawking describes it rather well:

> In other words, the more accurately you try to measure the position of the particle, the less accurately you can measure its speed, and vice versa. Heisenberg showed that the uncertainty in the position of the particle times the uncertainty in its velocity times the mass of the particle can never be smaller than a certain quantity, which is known as Planck's constant.[11]

But it goes even deeper than this as we will see in the next chapter.

In 1927, many of the leading physicists throughout Europe met together at a conference in Brussels to discuss the crisis. This was the fifth of what became the famous Solvay Conferences. Heisenberg said of that time, "we all stayed at the same hotel, and the fiercest arguments took place not in the conference but during hotel meals."[12] Hopefully they didn't resort to throwing their food.

On the agenda were the matrix mechanics of Heisenberg and his collaborators (Max Born, Pascual Jordan, Paul Dirac and Wolfgang Pauli), Schrödinger's wave mechanics, Born's probability distribution and the 'Uncertainty Principle'. To everyone's surprise, it was found that all four approaches were describing the same thing. From it a consistent picture of the atom emerged, which is still valid, and became known as Quantum Mechanics.

References

1. Heisenberg, Werner., 1963. *Physics and philosophy.* London: George Allen & Unwin. p 38.
2. ibid., p 39.

3. Fraze, G., Lillestøl, E., and Sellevåg, I. 1994. *The search for infinity.*
1. London: Michell Beazley. p 36.
4. Weinberg, Steven., 1993. *Dreams of a final theory.* London: Vintage. p 52.
5. Hey, Tony, and Walters, Patrick., 1987. *The quantum universe.* London: Cambridge University Press. p 5.
6. Weinberg, S., op. cit. p 55.
7. ibid., p 56.
8. ibid., p 57.
9. Heisenberg, W., op.cit. p 42.
10. ibid., p 43.
11. Hawking, Stephen, 1988. *A brief history of time.* London: Bantam. p 55.
12. Fraser, G., et al., op. cit., p 37.

Chapter Fourteen
Magic that's Real

The End of Certainty

The reason why the Uncertainty Principle is so important, is that it proves once and for all time that we will never be able to measure anything with absolute final accuracy, no matter how clever our technology becomes. This is because it is not a question of how good an experiment can be, or how perfect our instruments are; it is about Nature being immeasurable. Hawking puts it bluntly: "Heisenberg's uncertainty principle is a fundamental, inescapable property of the world." He goes on to say that it, "had profound implications for the way in which we view the world."[1]

This is one of the critical discoveries of the Quantum Revolution. It overturned all previous thinking. That's why it is the greatest revolution in human consciousness so far. Ever since Galileo and Newton, people had thought that one day it would be possible to measure everything exactly down to the last detail, and that we would then understand how the universe worked. Most people still think this way, and imagine that one day science will explain everything. It is a way of thinking that became known as Classical Determinism or just Determinism.

It's a perfectly sound idea because it's based on the laws of cause and effect, which we see around us every day. If you strike a pool ball; the cause – then, (if your aim is good) it will hit the black ball in such a way that it drops into the pocket; the effect, and you've won the game.

A famous example of this way of thinking comes from the mathematician and astronomer Simon de Laplace (1749 – 1827), who was so impressed by Newton's laws and their application to the planets that he made a leap of imagination into the future. He assumed that one day a super intelligence would be able to explain everything. Here's what he said.

> An intelligence that at a given instant was acquainted with all the forces by which nature is animated and with the state of the bodies of which it is composed would – if it were vast enough to submit these data to analysis – embrace in the same formula the movements of the largest bodies in the universe and those of the lightest atoms: Nothing would be uncertain for such an intelligence, and the future like the past would be present to its eyes.[2]

The story goes that when he presented his major treatise to the Emperor Napoleon I, the Emperor, after reading it, said to him, "but Sir, there is no mention of God in your treatise?" Laplace is said to have replied dismissively, "I had no need of that hypothesis"!

With the discovery of the Uncertainty Principle, scientists began to realise that Laplace's dream, and the dream of all of classical science, had passed away forever. Nothing could be absolutely certain again. As Hawking says, this changes the way humans must view the world. Because the Uncertainty Principle is an 'Absolute of Nature', we should not think of it as just something to do with science; it should become part of our system of values. Every dictator that ever was believed that he was absolutely right, as do most religions. It is an attitude that is responsible for more death and destruction than any other in the history of our species. It is surely part of humanity's process of growing up to realise that any certainty is suspect, and as we mature as a species we should surely learn to accept that uncertainty is universal and work with it. Only then will we acquire maturity.

On a personal basis one of the most tragic images of the twentieth century for me, is that of the Jewish scientist Bronowski in the BBC series 'The Ascent of Man', standing in a muddy ditch at Auschwitz and quoting Oliver Cromwell to the Nazis who murdered his family, "I beseech you, in the bowels of Christ, think it possible you may be mistaken."[3]

Baffled

What replaced Classical Determinism was the Quantum Revolution. The nature of the world, in fact the nature of 'reality', is full of uncertainty. Things that still don't add up a hundred years since they were first discovered. Even the pioneers who founded the theory were baffled by it. Niels Bohr is often quoted as saying, "If you are not shocked by quantum mechanics you have not understood it." Similarly, Schrödinger once said, "Had I known that we were not going to get rid of this damned quantum jumping, I never would have involved myself in this business."[4]

Richard Feynman, a Nobel Prize winner and one of the most outstanding scientists of the 20th Century, used to begin by telling his students that they would not understand quantum mechanics even after he had explained it to them. Here is what he says to the readers in one of his books.

> Why, then, am I going to bother you with all this? Why are you going to sit here all this time, when you won't be able to understand what I am going to say? It is my task to convince you *not* to turn away because you don't understand it. You see, my physics students don't understand it either. That is because *I* don't understand it. Nobody does.[5]

Steven Weinberg tells the story of how he once asked a colleague about a promising young student who had dropped out of sight. His colleague shook his head sadly and said, "He tried to understand quantum mechanics." Weinberg continues, "But I admit to some discomfort in working all my life in a theoretical framework that no-one fully understands."[6]

Don't worry too much about all this 'not understanding' business - I am confident we can come to the same level of understanding as most people.

THE FINAL MYSTERY

The Experiment with Two Holes

Fortunately, the central mystery of the quantum world can be summed up in one simple experiment, known as Young's double slit experiment. In 1804 Thomas Young devised a series of experiments which seemed to show conclusively that light must be a wave. It was a landmark discovery at the time because everyone had believed Newton who was sure that light was 'corpuscular' and came in particles. So Young's wave picture of light became the standard scientific view, until Einstein's paper on the photoelectric effect forced people to examine the wave-particle possibility. This is the experiment that Feynman always referred to as, "the experiment with the two holes".

What Young did was to set up a light source (an ordinary torch will do), and then place a barrier which had two narrow slits in it, in the path of the light. Behind this he placed a screen to detect what happened to the light. He found that the light passed through both holes and produced the characteristic fringe, or 'interference', pattern of light and dark lines, just like the ripples on the surface of a pond, which shows that light must be a wave.

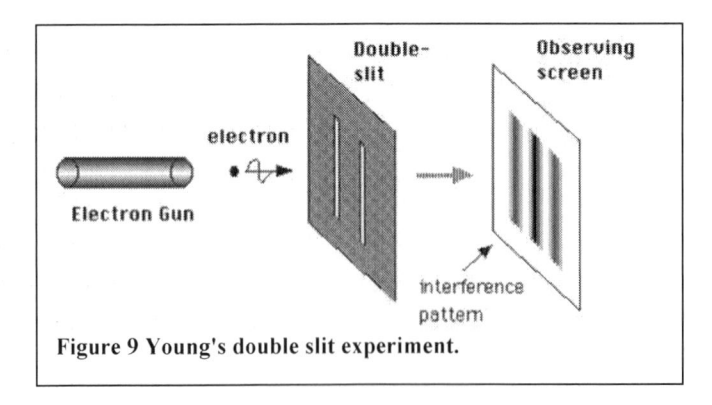

Figure 9 Young's double slit experiment.

However, when you then decide to block one of the holes, the light passes through the other and arrives on the detector screen as a fuzzy blob, and the interference pattern disappears, showing that light is a particle. Unblock the hole and the wave pattern returns, and doesn't coincide at all with where the blob was.

If, instead, you fired a rapid machine gun at a thick concrete wall with two narrow slits, with a target screen behind, some bullets would go through one hole and hit the target behind, whilst others would go through the second hole.

After firing for a while you would end up with two separate blobs behind each hole on the target. The same result would happen if you were shooting golf balls or throwing baseballs. But this doesn't happen in the world of the very small, because subatomic particles are not real in the same way that bullets or golf balls are. Yet everything is made of subatomic particles. Is our world therefore in some manner not quite real?

Is Everything Unreal?

You may be thinking, well, light is a pretty insubstantial thing anyway, so maybe it's just photons that behave in this weird way. Unfortunately, as de Broglie showed, all particles behave in the same way. If you change the photons for electrons or protons or neutrons, the same thing happens.

Let's go back to our experiment and, instead of photons, use electrons. Let's take a really careful look at what is happening, this time by firing just one electron at a time. "One would expect each to pass through one slit or the other", Hawking says. "In reality, however, even when the electrons are sent one at a time, the fringes still appear. Each electron, therefore, must be passing through *both* slits at the same time."[7] This is what is meant by being in two places at once. Yet if we block one of the slits again, then the wave fringe disappears. The big question is: how on earth does each electron know in advance what *we* have decided to do?

It becomes even stranger still when we place an electron detector right next to each hole, to measure the electrons as they pass.

When we do this the wave pattern disappears, even though both slits are still open. "But if we don't have instruments that can tell", Feynman says, "the interference effects come back! Very strange indeed!"[8]

Particles with Consciousness

It seems then, that whenever we try to detect the passing of an actual electron, it somehow knows we are looking, and defiantly behaves like a single particle. But when we stop looking or measuring, it is somehow aware that we have taken the detectors away, and it then continues going through both holes at once. It is actually, literally, in two places at once. This is absolutely astonishing when you think about it. But it gets even more exciting than this.

As I have said, the description of being in two places at once is called the "wave function". But it is a strange kind of wave. "It is spread out in principle, to fill the universe."[9] The whole universe? This is vividly illustrated by another version of the same experiment described by Paul Davies.

> Moreover, if instead of repeating the experiment, electron by electron, a whole collection of laboratories try it out independently, and they each pick just one photograph at random, then the collection of all these separate independent photographs also shows the interference pattern. These results are so astonishing that it is hard to digest their significance. It is as though some magic influence was dictating events in different laboratories…in conformity with some universal organising principle. How does any individual electron know what the other electrons, maybe in other parts of the world, are going to do?…How is their preference controlled at the individual level? Magic?[10]

There will be more about this 'magic' later.

References

1. Hawking, Stephen., 1988. *A brief history of time*. London: Bantam. p 55.
2. Layzer, David., 1990. *Cosmogenesis*. Oxford: Oxford University Press. p 5.

3. Bronowski, J., 1974. *The ascent of man.* London: BBC. p 374.
4. Gribbin, John., 1988. *In search of schrödinger's cat.* London: Corgi. p 117. 4.
5. Feynman, Richard., 1985. *QED.* Harmondsworth: Penguin. p 9.
6. Weinberg, Steven., 1993. *Dreams of a final theory.* London: Vintage. p 66.
7. Hawking, S., op. cit. p 59.
8. Feynman, R., op. cit. p 81.
9. Gribbin, John., 1986. In search of the big bang. London: Corgi. p 23.
10. Davies, Paul., 1988. Other worlds. Harmondsworth: Penguin. p 66.

Chapter Fifteen
Infinite Paths

Here and There

In the meantime, let's return to Earth a little. If that's indeed possible. To explain what is happening, physicists say that before a particle is actually detected or measured, it exists in a 'superposition' of a vast collection, actually an infinite number, of wave functions. They then calculate, using Schrödinger's famous wave equation, the area where the wave functions are most intense, which gives the greatest probability of finding the electron in that locality. At the instant that the electron is detected, all the trillions of other possibilities vanish. This is called the 'collapse of the wave function'. Using a very simple model, Steven Weinberg describes it like this:

> We can think of this system as a mythical particle with only two instead of an infinite number of possible positions – say *here* and *there*. The state of the system at any moment is then described by two numbers: the *here* and *there* values of the wave function...when we are not observing the particle, the state of the system could be pure *here*, in which case the *there* value of the wave function would vanish, or pure *there*, in which case the *here* value of the wave function would vanish, but it is also possible (and more usual) that neither value vanishes and that the particle is neither definitely *here* or definitely *there*. If we do look to see whether the particle is *here* or *there*, we of course find that it is in one or the other positions; the probability that it will turn out to be *here* is given by the square of the *here* value just before the measurement, and the probability that it is *there* is given by the square of the *there* value."[1]

Infinite Possibility

If you are feeling a bit dizzy, don't worry, so am I! Let's take a break for a minute and have a closer look at the actual make up of these waves. John Bell, the author of Bell's Inequality (which we will come to later), had this to say about quantum waves: "It is the mathematics of this wave motion, which somehow controls the electron." ... mathematics? Makes you feel as though you've just stepped into Alice's Wonderland. He continues with the question that we are looking at: "What is it that 'waves' in wave mechanics?" he asks. And then goes on like this, "in the case of water waves it is the surface of the water that waves. With sound waves the pressure of the air oscillates. Light also was held to be a wave motion in classical physics. We were already a little vague about what was waving in that case...and even about whether the question made sense. In the case of the waves of wave mechanics we have no idea what is waving". All that can be said, he continues, is "what we do have is a mathematical recipe for the propagation of the waves, and the rule that the probability of an electron being seen at a particular place when looked for there, is related to the intensity there of the wave motion."[2] So it looks very much as though maths is at the very heart of literally every – 'thing'. The mystery of what maths *is* begins to deepen. James Gleick, in a book about Richard Feynman, says, "Schrödinger's waves defied every conventional picture. They were not waves of substance or energy but of a kind of probability, rolling through a mathematical space."[3] Mathematical space?

Sir Roger Penrose, who is the Emeritus Rouse Ball Professor of Mathematics at Oxford University and a major international figure, says about quantum mechanics: "so even material substance seems able to convert itself into something with a more theoretical mathematical actuality. Furthermore, quantum theory seems to tell us that material particles are merely 'waves' of information."[4] I feel I should apologise for so many quotes all in a row, but if I didn't use these well known scientists you would certainly think I'd made it up!

So far we have found that quantum waves are largely mathematical objects and that they somehow contain an infinite number of possibilities until the moment of measurement or observation. It is as though every possible option for the particle actually exists before a measurement is made.

A famous mathematician, David Hilbert (1862 – 1943), whom we'll hear more about later, discovered what must be one of the strangest things for a human to try and imagine. It goes like this. We all live in a three dimensional world of up-down-and-sideways, with a fourth if you count Einstein's space-time. Our experience of the 3D world we live in is pretty much as fundamental as you can get. But the monster that Hilbert discovered is actually an 'infinite dimensional space' – a space with literally an infinite, an absolutely countless bazillion number of different dimensions! It's likely to be one of the best mind-benders of all time.

It's now called a 'Hilbert Space'. And to crown this fantastic beast, the 'wave function' we have been talking about actually lives within a Hilbert Space. So not only do electrons and protons and all the other atomic particles travel by an infinite number of paths – or by every *possible* route, they exist within an infinite dimensional space! There is still a further quality which is equally excruciating to the mind. It is that all the infinite possibilities actually contribute something 'real' to the outcome, (i.e. an electron being in a particular place). Weird though it may seem, it suggests that infinities can actually affect our real world of atoms, and atoms are what everything is made of. In fact, Professor David Deutsch of Oxford University, who is the founder and pioneer of Quantum Information Science, says that it is the infinite surrounding paths that make our real world possible. "The positively and negatively charged particles would simply move out of position and crash into each other, and the structure would disintegrate. It is only the strong quantum interference between the various paths taken by charged particles in parallel universes that prevents such catastrophes and makes solid matter possible."[5] Don't worry about the mention of parallel universes, we are coming to that shortly. I just wanted to show you that the infinite paths of the quantum superposition are what actually make our world 'real'.

Richard Feynman devised his own method for calculating the paths of subatomic particles, which he called the 'sum over histories' method

INFINITE PATHS

which has since become the standard method. In this model the particle goes through the double-slit experiment by *every possible* path. This is the equivalent of the infinite possibilities, but with the additional property of movement in time. Here is how he described it to another famous scientist, Freeman Dyson, whom we met earlier.

> The electron does anything it likes. It just goes in any direction at any speed, forward and backward in time, however it likes, and then you add up the amplitudes and it gives you the wave function.[6]

This is another extraordinary property of the quantum world. Notice that Feynman says the electron can go 'forward and backward in time'. According to Feynman's model, 'antimatter' (discovered by Dirac who also comes into our story shortly), is actually ordinary matter travelling backwards in time! John Cramer, Professor of Physics at the University of Washington in Seattle and others, have carried out experiments to actually test this. It's called retro-causality.

Feynman's explanation as to why light travels in a straight line also explains how photons know whether one or both slits are open and whether you have decided to use a detector.

> When all possible paths are considered, each crooked path has a nearby path of considerably less distance and therefore much less time...The nearby nearly straight paths - also make important contributions. So light doesn't *really* travel only in a straight line; it 'smells' the neighbouring paths around it.[7]

So it's the infinite paths that actually support the 'real' path, and therefore it is the infinite paths which make our world possible. It's as though these infinite alternatives are responsible for making our universe real.

The Cat in the Box Experiment

The strange nature of the quantum world worried many of those who discovered it. As I have said, Schrödinger was one of them. To show up the absurdity, the sheer craziness of what was being discovered, he devised a now famous thought experiment. No actual cat has ever been used to do this experiment.

What the experiment illustrates has become known as the 'Measurement Problem' which still rages to day. Here it is: before we make a measurement or observation of a quantum set up – say Young's experiment – each electron or particle is in the form of a vast number of individual wave functions, one for each possibility. As I just mentioned, this is known as the superposition.

When we make an observation, the particle suddenly stops being in all these possible states and all the waves collapse down to one – and the particle is found. The mystery of the Measurement Problem is: what makes the waves collapse? What is it about 'measuring' or 'observing' that transforms these tiny bits of atoms from being everywhere at once, to being suddenly in one place?

The natural thing to do would be to look at the detector first. This is usually an ordinary object like a Geiger counter. But because it is an ordinary object it is also made up of molecules and atoms and subatomic particles, which are in as much of a quantum state as the particles in the experiment. In this case a cat. So they can also be described by wave functions. In trying to find an answer to the Measurement Problem we have to admit that the detectors are part of the quantum world. So where does this chain of observation end? Perhaps it ends with us? But we are also made out of quantum objects. Is it then our consciousness that does it? Strange as it may seem, several important scientists have taken this position. They have suggested that human consciousness creates the world. But if it is us that collapses the wave function then we are faced with a puzzle. This is how Schrödinger first described his thought experiment in 1935.

A cat penned up in a steel chamber, along with the following diabolical devise (which must be secure against direct interference by the cat): in a Geiger counter there is a tiny bit of radioactive substance, so small, that perhaps in the course of one hour one of the atoms decays, but also with equal probability; perhaps none; if it happens, the counter tube discharges and through a relay releases a hammer which shatters a small flask of hydrocyanic acid. If one has left this entire system to itself for an hour, one would say that the cat lives if meanwhile no atom has decayed. The first atomic decay would have poisoned it.[8]

To our normal way of thinking this wouldn't be a problem. We would be certain that the poor cat was either dead or alive. But according to the rules of the quantum world the whole set-up within the box is in a superposition of many states, including one in which the cat is alive, and another in which it is dead. So it would have to be both dead and alive at the same time! It is only when we lift the lid on the box and look inside that the superposition of all the waves collapses and we find either a dead cat or a live cat.

Is it then our observation that makes the world real? Einstein couldn't agree. In a famous anecdote he is said to have asked, "surely the moon exists whether or not somebody is looking at it."

References

1. Weinberg. Steven., 1993. Dreams of a final theory. London: Vintage. p 59.
2. Bell, J.S., 1987. Speakable and unspeakable in quantum mechanics. Cambridge: Cambridge University Press. p 187.
3. Gleick, James., 1992. Genius, Richard Feynman and modern physics. London: Little Brown. p 75.
4. Penrose, Roger., 1994. Shadows of the mind. Oxford: Oxford University Press. p 14.
5. Deutsch, David. 1998. The fabric of reality. London: Penguin. p 213.

6. Davies, Paul., 1988. Other worlds. Harmondsworth: Penguin. p 66.
7. Feynman. Richard., 1985. QED. Harmondsworth: Penguin. p 9.
8. Davies, P., and Brown, J. edited. 1986. The ghost in the atom. Cambridge: Cambridge University Press p 29.

Chapter Sixteen
The Great Smoky Dragon

The Measurement Problem

Strange as it may seem, the simple cat in the box paradox lies at the centre of a huge controversy that has lasted from the 1920's to the present day. Whilst everyone agrees that quantum theory is the most successful theory in the history of science, no one is sure what it means! Although the Quantum Revolution has given us our plasma screens, GPS, iPhones, laptops and so on, no one can agree about what happens when we actually come to measure a bit of an atom, say, an electron or proton.

The central dilemma is that when these bits are not being measured or observed by anyone, they clearly behave like waves, but as we have seen - not waves of *anything* in particular. The closest we can get is that they are waves of mathematical probability until the instant of observation or measurement. Before this, they exist in a superposition of trillions upon trillions of possibilities. John Wheeler, one of the major figures of 20th century physics who took part in the development of both the atom and hydrogen bombs, and was the man who first coined the phrase 'black hole' to describe a space-time singularity, called this strange mystery: 'The Great Smoky Dragon'.

> The amazing feature of the elementary quantum phenomenon - the Great Smoky Dragon - is exactly this. It is indeed something of a pure knowledge – theoretical character, an atom of information which has no localization in between the point of entry and the point of registration.[1]

The Practical View

Needless to say, the great majority of working physicists don't give much of their time to wondering about the Great Smoky Dragon. They simply apply the necessary mathematical recipe and get on with the job, knowing that at some point between the micro level of subatomic bits, and the macro level 'click' of a Geiger counter, quantum events somehow manage to 'turn into' our ordinary everyday reality. Someone once said that the average physicist using quantum mechanics thinks less about the Great Smoky Dragon than a car mechanic.

However, there are others who want to know what is going on between the head and the tail of the dragon. Not least because the whole world – the whole universe never mind - is made out of atoms, including our own minds. Over the years four main explanations of the 'Great Smoky Dragon' have arisen. Each of them is quite alarming because they are not at all what you would expect, especially as this is the very frontier of human knowledge. One of them neatly avoids saying anything, but the other three are far stranger than fiction.

The Orthodox View

The first way of explaining the Smoky Dragon is the least interesting, but until quite recently it was the accepted view, which most physicists followed. It is known as the Copenhagen Interpretation after its Danish founder, Niels Bohr. What he said, in a very long-winded way, was in essence that science cannot say anything about the great Smoky Dragon. You can specify in detail your experiment, say, for example, Young's double slit; and describe very clearly what source of particles you are going to use, and the exact nature of your detectors. Then when you have carried out the experiment you record the precise results. But unfortunately you are not allowed to say anything about what happens in between, because nothing physical can be said about it.

Bohr wrote laboriously and boringly about a notion he called 'complementarity'. If you don't know what that means you are not alone, many famous scientists didn't know either, (nor does Windows spell-check). It was an attempt at saying that a particle can posses a position in space or it can have a momentum, and that these two properties are complementary, even though, because of Heisenberg's Uncertainty Principle, the particle couldn't have both properties at the same instant. In the early 1900's there were a group of thinkers in Vienna called the Positivists who refused to believe in the existence of atoms because they couldn't see them. It sounds to me a bit like that. Since he didn't know what to make of the Smoky Dragon, Bohr appeared to side-step the issue by saying that nothing can be said about what happens between the particle leaving and its arrival at the screen.

This became the standard interpretation. It was basically the most convenient way of getting round the problem, and so most people accepted it. I always felt it was a bit of a cop-out. Its main worth, to my way of thinking, was that it advertised just how completely puzzled people were about what was going on - so puzzled that they had to contrive an excuse not to ask.

I was pleasantly surprised then when I found that David Deutsch agrees. "Thus generations of physicists have found it sufficient to regard interference processes...as a 'black box': they prepare an input and observe an output. They use the equations of quantum theory to predict the one from the other, but they neither know or care *how the output comes about* as a result of the input."[2] (His emphasis.)

About the Copenhagen Interpretation he says, "As far as the unobserved events that interpolated between conscious observations, one was 'not permitted to ask' about them! How physicists...could accept such an insubstantial construction as the orthodox version of a fundamental theory is a question for historians."[3]

Humanity Takes Centre Stage

The second interpretation is in some ways quite straightforward and seems to be most obvious, but it leads to results that have monumental implications which are difficult to come to terms with. It is known variously as 'observer participation' or 'mind over matter'.

It was first put forward by the Hungarian-born American physicist Eugene Wigner who was a leading figure in the development of quantum mechanics and won the Nobel Prize in 1963. According to Wigner, it is human consciousness that makes the quantum wave collapse from being in a state of an infinite number of possibilities to just one, which is where the particle is found. In other words, the cat in the box remains both dead and alive until a human consciousness opens the lid of the box and observes it.

This is how Paul Davies explains it: "In Wigner's interpretation of quantum theory, the minds of sentient beings occupy a central role in the laws of nature and in the organisation of the universe, for it is precisely when the information about an observation enters the consciousness of an observer that the superposition of waves actually collapses into reality. Thus in a sense, the whole cosmic panorama is generated by its own inhabitants! According to Wigner's theory, before there was intelligent life, the universe did not 'really' exist."[4]

Bizarre? Unfortunately Wigner was no ordinary scientist. Apart from the Nobel Prize, he worked on the Manhattan Project to make the first atom bomb, was Director of Research at the Clinton Laboratories before becoming Professor of Mathematical Physics at no less a university than Princeton. His other awards include the U.S. Medal for Merit (1946), the Enrico Fermi Prize (1958), Atoms for Peace Award (1960), the Medal of the Franklin Society, the Max Planck Medal, the U.S. National Medal of Science, and honorary doctorates from some *nineteen* universities worldwide. He was President of the American Physical Society and a member of the National Academy of Science, as well as a member of the Royal Society in the U.K. He was what you might call a major heavyweight of the scientific establishment, so his ideas are taken seriously.

It still gives me the shivers. If Wigner's theory is correct, then human beings are somehow responsible for the universe being real. Perhaps when no one is looking at the moon it does revert to being just a presence of overlapping possibilities? Just as the sun seems to go around the Earth, perhaps in a way which we don't yet understand, our consciousness may be creating reality. If so, then we are certainly centre stage, and not just an insignificant collection of carbon atoms lost in a gigantic cosmos.

Inventing a Pilot

The third way of understanding the Great Smoky Dragon is known as the de Broglie – Bohm pilot wave or quantum potential theory. In this version the quantum wave is not just a mathematical device; it is an actual wave of potential, something like an electromagnetic wave which guides each particle through one or the other slit in the experiment. The difference is that in order for the theory to work, the wave has to be everywhere at once, or what is called 'non-local'.

When quantum mechanics was first being developed, many scientists (Einstein included) hated the idea that you couldn't pin down both the position and the speed of a particle. It wasn't science as they knew it. Science was nothing, in their view, if it couldn't be predictable. So they were sure that there must be something within the wave that was briefly robbing the particles of their position and momentum, and that given time it would be discovered. They called this something a 'hidden variable' and set about inventing ways in which it might work.

In 1927 Louis de Broglie, the French prince who first suggested that electrons might also be waves, came up with a solution. He invented what was called a pilot wave that somehow guided or piloted each particle, by letting it know if both, or just one, of the slits was open. His model was promising and seemed to work. Unfortunately, in 1932 a prominent international mathematician called John Von Neumann constructed a proof which seemed to show that all hidden variable theories like de Broglie's were doomed never to work properly.

Von Neumann was a Hungarian born in Budapest, the son of a banker. He was a mathematical prodigy as a child, and was educated at the best universities in Budapest, Berlin, and once again, Göttingen (Planck & Heisenberg). Because he was Jewish he moved to America in 1930 to avoid the Nazis, and later worked on the Manhattan Project to produce the atomic bomb. Together with J.L. Tuck he developed the high explosive lens that was essential to the actual detonation of the bomb. He is best known as a pioneer in the development of electronic computers. On top of this he made many original contributions to mathematics, including game theory which is important in artificial intelligence circles.

His reputation, even in 1932, was such that when his book "Mathematical Foundations of Quantum Mechanics" came out, debunking all hidden variable theories, everyone jumped on the bandwagon and, without thinking, rejected de Broglie's model out of hand. John Bell describes how de Broglie "was laughed out of court in a way that I now regard as disgraceful, because his arguments were not refuted, they were simply trampled on."[5]

Steven Weinberg seems unaware of de Broglie's humiliating treatment and suggests that he could have contributed more. "With such a doctoral thesis behind him, it might have been expected that de Broglie would go on to solve all the problems of physics. In fact he did virtually nothing else of scientific importance throughout his life."[6]

I can't help thinking that if I were a bona fide aristocrat of independent means, I might also have abandoned science to its own devices.

Sweet Revenge

Much later, it was John Bell who took the trouble to look carefully at de Broglie's work and found it to be largely consistent. Naturally this prompted him to then examine Von Neumann's proof. Euan Squires of Durham University points out, "Although several people seemed vaguely to have realised this problem with Von Neumann's theorem, it was not until 1964 that John Bell finally clarified the issue, and removed this theoretical obstacle to hidden-variable theories. The article was published in 'Reviews of Modern Physics', 38, 447 (1966)."[7]

John Bell was understandably not quite so polite about it. After all, a man who might have contributed much to science had had his reputation ruined by a heavyweight, who it turned out, had made an embarrassingly stupid mistake!

> The Von Neumann proof, if you actually come to grips with it, falls apart in your hands! There is *nothing* to it. It's not just flawed, it's silly! When you translate (his assumptions) into terms of physical disposition, they're nonsense. You may quote me on that: The proof of Von Neumann is not merely false but *foolish*!"[8] (his emphasis.)

Rebel with a Cause

The pilot wave, 'quantum potential' version of the Great Smoky Dragon has another human story to tell. It concerns the counter-revolutionary David Bohm. He began work as a student during the Second World War at Berkeley in California, under Robert Oppenheimer who was in overall charge of the Manhattan Project, and he contributed to the project himself in a small way.

This was now some twenty years after the Copenhagen Interpretation had become the orthodox view. After the war, Bohm moved to Princeton University to become a young member of the physics faculty, and while tinkering with writing a text book, he found himself rejecting the conventional approach and sympathising with de Broglie's pilot wave idea. This led him to propose the heretical alternative that every quantum particle was in fact under the influence of a hidden variable that he called the 'quantum potential'. It was this potential that linked all the particles together in an interconnected whole so that each particle could tell instantly what the others were up to.

The uncertainty that had bothered everyone so much was seen to be the result of fluctuations or changes in this quantum potential. He was even able to show that his theory could lead automatically to Heisenberg's Uncertainty Principle. But because it was revolutionary and went against the established orthodoxy, it was largely ignored. This often seems to happen in science.

Bohm showed it to Einstein, who by then had settled at the Institute for Advanced Study near Princeton. He was quite sympathetic, but disliked the theory because it was non-local. In order to work, the quantum potential had to be assumed to fill the whole universe, allowing instantaneous communication between particles that could be light years apart. This went against Einstein's Special Relativity and its light speed barrier. In an interview Bell said of this aspect of the theory, "but it did have the remarkable feature of action-at-a-distance. You could see in the equations of that theory that when something happened at one point, there were consequences immediately over the whole of space, unrestricted by the velocity of light."[9]

Years later, in 1982, an experiment was carried out that proved beyond doubt that action-at-a-distance is a part of Nature. We will be investigating this in greater depth. Many other similar experiments have since confirmed this phenomenon, and Bohm's theory has not been contradicted. However, at the time he was ignored. The story goes that some students at one of the famous universities took Bohm's theory to their professor who, after looking at it, couldn't find anything wrong with it. Finally they took it to the great Robert Oppenheimer himself who pronounced that it was well known that Bohm's ideas were wrong – but the students persisted. Finally Oppenheimer is supposed to have said: "Well, we can't find anything wrong with it, so we'll just have to ignore it."[10] Sounds a bit brutal and arrogant.

Not long after this the Cold War between the West and the USSR began, including incidents like the Berlin airlift when the Soviets blocked all roads into the city. At the time, the American authorities became somewhat paranoid about communist infiltrators giving atomic secrets to the Russians. Bohm was called before the un-American Activities Committee at the House of Representatives, and asked to testify about the politics of some of the scientists he had known when working on the Manhattan Project.

Bohm refused on principle to testify about the personal lives of his colleagues, and two years later he was indicted for contempt of Congress, and brought to trial. He was acquitted, but it was now the

McCarthy era in the US and, despite being a brilliant theorist, the walls of academia closed ranks and he was refused a university post in the U.S. After travelling a bit, he found a haven in the mother of all democracies, and settled at Birbeck College in London where he continued to develop his theory, becoming internationally respected until his early death in 1990.

To re-cap so far on the Great Smoky Dragon. The first, and orthodox, view is that nothing can be said about a particle when it is not being observed or measured. The second view, put forward by Wigner, is that the particle remains in a state of many possibilities, and that it is human consciousness that collapses the wave function and makes it 'real'.

The third version is that the particle is not a mathematical probability but a real part of a wave of potential. But this leads to the equally counter-intuitive conclusion that everything in the entire universe is somehow connected. In other words, at the quantum level the universe is an undivided whole, in which each particle somehow knows what every other particle is doing.

Many Worlds

The fourth version is known as the 'Many Worlds' or 'Multiple Universe' theory, which was first put forward in 1957 by Hugh Everett when he was a student at Princeton being supervised by John Wheeler. It was first published in Reviews of Modern Physics, and because of its revolutionary nature David Deutsch recalls how, "Wheeler was afraid that Everett's idea might not be sufficiently appreciated. So he wrote a short paper to accompany the one that Everett published."[11]

In this version of the Smoky Dragon, all the infinite possibilities of the quantum wave are treated as 'real'. Yes, real! This does away with the need for the wave function to 'collapse'. The many possibilities don't collapse down to one probability, because all the possibilities are actually treated as fully real. In other words the cat is alive in one world and dead in another. Like some strange hallucination. What it means is that when we observe or measure a particle, we split into an infinite number of copies of our self, each one occupying a different world – in fact a different universe! I imagine if you tried to persuade a psychiatrist he would be convinced you had serious delusions and would be quietly signing the admission forms so you didn't notice.

Many people have heard of the 'many worlds' theory but most don't appreciate that it is actually the most literal interpretation of the quantum world. In other words, the most 'scientific' if you like. It doesn't say that the wave is a vast collection of ghostly mathematical possibilities; it actually treats each possibility as existing for real. Remember now, that the number of possibilities or paths, or alternate worlds if you like, is actually infinite. This is important for later because in this version the infinities are considered to be real! So real we can actually make use of them. If that isn't hugely exciting I don't know what is.

The result is an utterly astonishing view of reality. It seems that every instant of our lives we are being split into trillions of copies, along with everything else around us, and that existence itself is actually made up of an infinite number of parallel worlds of which we are not aware.

If you feel like giving up at this point I don't blame you, but before we go any further I should tell you that at many universities worldwide, as well as at big private laboratories like IBM, there is a race on to produce the first commercial quantum computer. A computer that will be able to use the alternate worlds which surround us for processing information in parallel on a scale that may be many thousands of times faster than now. Because of the existence of an infinity of parallel worlds, the hope is to be able to process information on an immense scale, so immense that conventional digital super computers would become prehistoric by comparison. David Deutsch says that as far back as 1989, "at IBM Research, Yorktown Heights, New York, in the office of the theoretician Charles Bennett, the first working quantum computer was built."[12] As I write, a company called D-Wave of Burnaby in Canada has sold the first quantum computer to Lockheed Martin, the giant aircraft maker, and more recently they sold another to Google. In some tests the newest model, D-Wave Two, was able to solve some problems 3 600 times faster than a conventional computer.

There will be more about this in part five. It brings to mind what J.B.S. Haldane, a famous Scottish scientist, once said, "the universe is not only stranger than we imagine, it is stranger than we *can* imagine." We have seen a number of examples now where Nature has proved to be much greater and more ingenious than human imagination. The multiple universes that surround us every fraction of a second of every day are a good example. At an international physics conference several years ago now, a group of seventy-two prominent physicists was asked whether they supported the Multiple Universe interpretation of quantum theory. Fifty-eight percent said they did, and amongst them were four Nobel laureates!

David Deutsch reminds his readers of how an eminent physicist called Bryce de Witt initially opposed Everett, and in an historic exchange of letters, put forward a number of objections to his theory, each of which Everett successfully rebutted. In the end de Witt admitted that he just couldn't feel himself 'split' into multiple distinct copies every time a decision was made. Everett's reply echoed the dispute between Galileo and the Inquisition. "Do you feel the earth move?" he asked, "de Witt conceded."[13]

Before he had received his Ph.D., Everett had already left Princeton University to work for the Pentagon as a defence analyst. Much of this work was a state secret and classified, but it is known that he worked on missile systems like the Minuteman Project. He later left the Pentagon and set up his own private company, becoming a multi-millionaire. He died of a heart attack aged 51 in 1982.

References

1. Davies, P., and Brown, J., Edited by. 1986. The Ghost in the atom. Cambridge: Cambridge University Press. p 66.
2. Deutsch, David., 1997. The fabric of reality. Harmondsworth: Penguin. p 330
3. ibid, p 328.

4. Davies, Paul., 1988. Other worlds. Harmondsworth: Penguin. p 132.
5. Davies, P., and Brown, J., op. cit. p 56.
6. Weinberg, Steven., 1993. Dreams of a final theory. London: Vintage. p 55.
7. Squires, Euan., 1986. The mystery of the quantum world. Bristol: Adam Hilger. p 78.
8. Gribbin, John., 1995. Schrödinger's kittens. London: Weidenfeld & Nicholson. p 155.
9. Davies, P., and Brown, J., op. cit., p 56.
10. Matthews, Robert., 1992. Unravelling the mind of god. London: Virgin. p 146.
11. Deutsch, David., op. cit. p 328.
12. ibid., p 218.
13. Ibid., p 328.

Part Four

Where Worlds Meet

Chapter Seventeen
Atomic Forces

How it started

The first person to take the Big Bang idea seriously was, as we said earlier, George Gamow. He wondered where all the ninety-two elements like oxygen, carbon, iron and uranium had come from. What he suspected was that they had been forged in the first few moments when the universe was born.

His key realisation was that the study of the universe, which we now call cosmology, might also help explain the world of the very small, in other words, the world of atoms and visa versa. This was a breathtaking vision at the time because it promised to unite the study of everything. Although he was wrong about almost all the elements, his line of thought became a major theme that ended up uniting astronomy and physics. As a result we now have scientists called astrophysicists.

He was also responsible for a well-known publishing joke. He had a Ph.D. student named Ralph Alpher, whom we met earlier and, to draw attention to his student's work, he contacted a friend, Hans Bethe, who was also a physicist, and invited him to put his name to a paper about the Big Bang, along with Alpher and Gamow. It appeared in the April Fool's Day edition of the learned journal 'Physical Review' in 1948 as co-authored by Alpher, Bethe and Gamow. Thereafter it was always known from the first three letters of the Greek alphabet as the 'Alpha Beta Gamma paper' – quite appropriate for a paper about the beginning of everything.

As time went on there was more and more co-operation between physics and cosmology. In the early days it began with high altitude balloons to do experiments with cosmic rays, which are bits of atoms that bombard the earth from outer space, and later with the construction of the biggest

THE FINAL MYSTERY

machines on earth like the Large Hadron Collider which is able to reproduce in miniature the conditions that existed fractions of a second after the Big Bang.

Discovering the Neutron

We left the story about discovering the different bits of atoms back with Rutherford finding the proton in chapter twelve. So how was the neutron discovered? Rutherford knew he was still facing a puzzle because protons only made up half the weight of the nucleus. He also knew that the missing particle had to be electrically neutral because the charge of the negative electrons balanced the charge of the positive protons. Although it hadn't been seen yet, he decided to call it the Neutron or 'neutral one'.

It was his senior research assistant, James Chadwick, who set out to find it. Chadwick had a nasty experience during the 1st World War. He had the misfortune to be studying as a student in Berlin in 1914 when hostilities broke out, and he ended up being interned as an alien, and having to live in a racehorse stable for four years!
Finding the neutron proved a long haul but after twelve years, in 1932, he finally managed. This showed the atom as having a number of outer electron levels in orbit, so to speak, and a heavy centre or nucleus made up of protons and neutrons. It's the picture which has survived to the present. However, there were some surprises in store.

The First Big Link

The pioneer who first linked the world of the universe with the world of the atom was Paul Dirac. He was one of the most brilliant theoretical physicists of the twentieth century. He started out modestly by doing electrical engineering at Bristol University in the U.K., then later moved to Cambridge to do maths. In his career he taught in the U.S.A. and visited Japan and Siberia, which must have been quite unusual in those days. In 1932 he was appointed to the Lucasian Professorship in Mathematics at Cambridge, the same post that Isaac Newton held, and the one Stephen Hawking also held.

ATOMIC FORCES

Figure 10 Paul Dirac.

When he was still a student at Cambridge in 1925, he heard Heisenberg giving a talk on quantum mechanics, which he had not heard of at the time. When Heisenberg returned to Germany, he left a copy of an unpublished paper with Dirac's supervisor, Ralph Fowler, who showed it to Dirac. The paper was about Heisenberg's matrix mechanics, but it contained a problem which Heisenberg could not resolve. Dirac was quickly able to see the difficulty and get around it by using a branch of mathematics invented by William Hamilton a hundred years earlier and which, once again, had nothing whatever to do with physics. If this sounds a little familiar, I should warn you that perhaps the most spectacular example of the power of maths is waiting in the wings. By using the maths discovered by Hamilton, Dirac was able to prove that the equations of ordinary Newtonian mechanics were a special case of Heisenberg's quantum equations. That was a shock. It turned out that if you ignored the tiny Planck scale, you automatically got the ordinary classical mechanics of Newton and our everyday big world of pulleys and levers.

As if that wasn't enough, Dirac next found an entirely new branch of maths which he called quantum algebra. He used it to show that Heisenberg's matrix mechanics was the same thing as Schrödinger's waves! Fowler immediately recognised the importance of his student's work, and at his instigation it was published in the Proceedings of the Royal Society for December 1925.

It was this sort of brilliance that led Einstein, who you may remember did not consider himself a good mathematician, to say, "I have trouble with Dirac. This balancing on the dizzying path between genius and madness is awful."[1] Not that Dirac was ever considered mad. He married Eugene Wigner's sister and led a normal life.

Having discovered how to unite Heisenberg and Schrödinger's work with ordinary physics, Dirac went on to use his quantum algebra to combine the world of the very big with the world of the very small. What he did was to unite Quantum Theory with Einstein's Special Relativity. Remember that this was not Einstein's masterpiece the General Theory of Relativity. The attempt to join Quantum Theory with General Relativity still goes on. The current front runner to achieve this unification is Super String or M-Theory as we mentioned in chapter ten.

Never Before Imagined

It was while he was doing this work that Dirac made a really astonishing discovery, quite by accident. In trying to describe the behaviour of an electron, he found that the equations he was working with actually contained two different solutions. One corresponded to the usual negatively charged electron, and the other to an unknown positively charged particle of the same mass.

Like Einstein before him, Dirac had discovered something deeply profound about the universe which no one had asked for or even remotely suspected. Purely by using a system of arbitrary symbols – just squiggles on a piece of paper, he had made one of the greatest discoveries in all of science. A discovery that changed forever our understanding of reality. He had discovered the existence of antimatter! After this momentous event, it was found that every atomic particle has an antimatter partner and antimatter is now regularly created in accelerator experiments. But it goes far deeper even than this. It goes to the very heart of the moment of creation.

Whenever matter and antimatter meet up, they annihilate each other in a flash of gamma rays. At the creation event it is thought that there were nearly equal amounts of matter and antimatter and that they almost cancelled each other out, but not quite. Had this been any different then certainly our consciousness could never have evolved. What was left over became our universe of matter. What Dirac discovered at his desk is absolutely crucial to the way our reality is constructed, going right back to billionths of a second after the Big Bang. What the equations had produced was the anti-particle to the electron that was later called the 'positron' or positive electron.

Dirac published his results in 1929 and three years later Carl Anderson in the U.S.A. was using a cloud chamber to analyse cosmic ray particles when he picked up the tracks of positrons. The maths had been right. This is perhaps the most dramatic confirmation of the mysterious power of mathematics to describe our physical world.

In 1933, a year after Anderson's discovery, Dirac received the Nobel Prize. Although not well known outside academic circles, Dirac is actually considered by most physicists to be one of the greatest in their field, able to compare favourably with the really big names like Newton, Maxwell and Einstein. His contemporaries appreciated his greatness and buried him in Westminster Abbey next to Newton. He was notoriously reserved, which is illustrated by an anecdote told by Heisenberg, who was once travelling with him on board ship from the U.S.A. to Japan.

In the evening on board ship, dances were held for the passengers, which Heisenberg indulged in while Dirac sat watching. "As Heisenberg came back to his chair after a dance, Dirac asked him: 'Why do you dance?' Heisenberg replied, 'Well, when there are some nice girls it is a pleasure to dance.' Dirac thought about this for a while; then after about five minutes, he said, 'Heisenberg, how do you know beforehand that the girls are nice?'"[2]

The Smallest Thing in the Universe

The story of the smallest thing in the universe goes back to Henri Becquerel and his lucky accident when he wrapped uranium salts together with a photographic plate and left them in a dark drawer. Uranium is a massive atom, in fact it's the biggest. It has 92 protons and 146 neutrons in the nucleus, and 92 electrons zinging around outside in different orbits. This makes it pretty cumbersome and a little shaky, so

THE FINAL MYSTERY

bits easily fly off in a process called beta decay. Left alone uranium eventually decays into lead which has a much more stable atomic structure. Interestingly, it is 70% heavier than lead which is probably why, as I said earlier, they use depleted uranium in tank and artillery shells.

The first person to work out what was happening to uranium was an Austrian-Swiss scientist called Wolfgang Pauli. He was one of the original pioneers of Quantum Theory, and is best known for the Pauli Exclusion Principle, which explained why electrons in an atom had to stay stuck in fixed orbits around the nucleus, and why only one electron per orbit is allowed in Nature.

In 1930 he found that in beta decay, which is what we now call radioactivity, there was a tiny amount of energy missing in the equations. He guessed that this must be in the form of an extremely tiny particle. Then in 1933 a larger than life Italian physicist, Enrico Fermi was able to explain the tiny missing mass. It was produced by a neutron turning itself into a proton. When this happens it spits out an electron plus a truly minute particle which he called a neutrino, or 'little neutral one'.

It sounds hardly believable, but these little bits are so small that they can easily pass right through the Earth and out the other side without bumping into a single bit of another atom on the way. In fact, there are probably thousands of them passing straight through your body at this moment. It is helpful here to remember that atoms themselves are vastly empty. Someone once worked out that it would be possible for a neutrino to pass straight through many miles of lead like a bullet through a cloud.

The whole idea is nothing but a mad assault on the imagination. So how do they know that neutrinos are real? How can you catch something that passes straight through everything else? They were sure they existed because of the bits left behind after they were created. But experimenters had to wait until the first nuclear reactors producing electricity were built in the 1950's in the U.S.A. and Britain, before they could catch any.

Literally billions of neutrinos are produced in the core of a nuclear reactor, and with so many being produced in one spot it was possible over a period of time to detect the presence of a few unlucky ones that bumped into detectors on their way out from the core.

The Weak Nuclear Force

In working out what happened in beta decay, Fermi had discovered an entirely new force in Nature, known as the weak nuclear interaction or 'the weak nuclear force'. This doesn't sound very important at first, until you realise that by discovering the weak nuclear force, science had raised the known number of forces in the universe from two to three – a big jump when you are talking about fundamental forces.

In Newton's time there were only two known forces: gravity and magnetism. People suspected that lightning was also a force, but it wasn't until Franklin and Faraday worked out what electricity was, that the known number of forces became three. The first great reduction in the number of forces came when James Maxwell showed that electricity and magnetism were just different aspects of one and the same force, which became known as electromagnetism. It was a great simplification. But with the discovery of the weak nuclear force the number once again increased to three:

(i) Gravity
(ii) Electromagnetism
(iii) Weak nuclear force

Because science seeks a single explanation for the universe physicists wondered if electromagnetism might be just part of the same force as the weak nuclear force. This was to become a major area of research over the next decades.

Enrico was one of the heroes of twentieth century physics. Born in Italy, the son of a railway official, he went first to the University of Pisa, and then worked under Max Born in Gottingen (again), before becoming a Professor at Rome in 1927. However, with the rise of Mussolini and the fascist party, and because his wife was Jewish, he emigrated to the U.S.A.

At Columbia University he exposed uranium to a bombardment of neutrons, and realised that this could produce a chain reaction that would release a huge amount of energy. Along with Einstein and Leo Szilard, he wrote to President Roosevelt about the dangers of Nazi Germany producing an atomic bomb first. As a result of their intervention, the President authorised the go-ahead for the Manhattan Project, which built the first atomic bomb.

Working on the top-secret project in Chicago, Fermi's group produced the first ever controlled self-sustaining nuclear reaction on 2nd December 1942. This proved that an atomic bomb was possible. The managing committee of the project was stationed at Harvard University, and the success of the experiment was passed to them in code. It read, 'The Italian navigator has landed in the New World'.

The Strong Nuclear Force

One of the major problems facing the model of an atom in the early 1930's concerned the protons. They knew each of them carried a positive charge that was equal to the negative charge of the electron. But this meant that the positive charge of the protons, clumped together in the centre of the atom, should make them repel each other, just like the same ends of a magnet repel each other. So why didn't all the protons in the nucleus of the atom simply fly apart?

Hideki Yukawa was a 28 year-old lecturer at Osaka University in Japan. One night in 1934 he couldn't sleep for thinking about the problem. In a flash of inspiration he realised that the protons must be held together by a really strong binding force – but the force had to work the opposite way to gravity and electromagnetism. With gravity, the further away two objects are, the weaker the force pulling them together. Perhaps the new force was the opposite of this and had less power the closer together the two particles were. Then, as you tried to pull them apart, the greater the force between them would become. In other words, as soon as they moved from zero distance apart, the force grew stronger and stronger.

Just like photons carry the electromagnetic force, Yukawa came up with the idea that the new force was carried by a messenger particle that was constantly zipping between protons so that they could never escape. His calculations showed that the particle had to be in between the weight of an electron and a proton, so he called it the middle one or 'meson'. He worked out that the meson should be two hundred times the mass of the electron, and so the new force would be much stronger than electromagnetism, and consequently it could hold the protons together.

In 1947 a British team led by Cecil Frank Powell discovered Yukawa's meson in a cosmic ray experiment, and it became known as the pi-meson, or pion for short. It was yet another triumph for maths. The new force that Yukawa discovered was called the 'strong nuclear force' and, in 1949, he became the first Japanese to win a Nobel Prize for Physics.

This meant that the number of fundamental forces had now increased to four:

(i) Gravity
(ii) Electromagnetism
(iii) The weak nuclear force
(iv) The strong nuclear force

The story of twentieth century physics is largely the story of how to unify these four forces to provide the ultimate prize of uniting them.

Why Stars Shine

It was only after physicists, working in the world of the very small, had discovered how the atom and its nucleus behaved that astronomers studying the big world began to understand how stars like our sun are able to shine.

In 1939 it was Hans Bethe who calculated how hydrogen burned into helium. This is called a thermonuclear reaction or a fusion reaction, in which hydrogen, with one proton in its nucleus, combines with other protons and neutrons to form helium, with two protons and two neutrons in its nucleus. When this happens the strong nuclear force binding protons together is broken and a large amount of energy is released, as in a hydrogen bomb.

This is the energy that streams 92 million miles (147 million kilometres) across space to warm our planet. Our sun, which began shining about 4 600 million years ago, burns something like 1 000 million tons of hydrogen into helium each second. The temperature required for this fusion to take place is about 10 million degrees centigrade. However, the sun is not about to cool down and fizzle out. It is estimated to last at least another 4 000 million years before it expands to become a red giant star that will engulf the Earth. Modern humans have only been around for at most 500 000 years. That's just a half of one million years, let alone 4 000 million. If we don't manage to destroy each other or the planet, then the human species will almost certainly have colonised other planetary systems by then. If you hadn't realised it before, it seems that we currently reside within the opening sentence of the epic that will become the story of humankind.

References

1. Hey, T., and Walters, P., 1987. *The quantum universe.* Cambridge: Cambridge University Press. p 123.
2. ibid., p 123.
3. Morris, Richard, *The Edges of Science*, Fourth Estate, London, 1992, page 24
4. Gribbin, John, *In Search of Schrödinger's Cat*, Corgi, London, 1988, page 201
5. Davies, Paul, *Superforce*, Unwin, London, 1990, page 105

Chapter Eighteen
The Standard Model

Infinities Again

Ever since Dirac's amazing work, there had been problems with the quantum version of electromagnetism. When theorists came to work out the exact charge on the electron, the answers kept giving them infinities. There seemed to be no way around the problem.

This is because infinities are the death of any kind of definite prediction. You can't say anything positive about the world if your information leads to an infinite amount. But we should bear in mind that neither can you say that these infinities don't exist. They are part and parcel of the same maths that so dramatically describes the real world. This strongly suggests that infinities are just as real as the rest of mathematics. It's just that you have to find a way of avoiding them if you want to work out how our particular 'slice' of reality works. The problem for physicists, as we will see later, is that maths is so vast it can describe every possible reality. Choosing which is our familiar slice is the difficult bit.

The answer to the exact charge on the electron was discovered independently by three people. The first was Julian Schwinger who had been a child prodigy, becoming a professor at Harvard at the age of twenty-seven. The second was Richard Feynman whom we've met before. He worked on the Manhattan Project during the war, and later became one of the most famous scientists of his age. He was a very colourful character. In his biography he relates how he took up bongo drums and ended up playing in a band in Brazil. He also experimented with cannabis and described in some detail a foolproof technique he developed for picking up girls in bars. Another expertise he taught himself was picking locks. He once moved some highly classified documents at Los Alamos from one top security safe to another and received a stern reprimand from Oppenheimer, the leader of the Manhattan Project.

The third person was Sin-Itiro Tomanaga, working in isolation in war-torn Japan. What the three came up with was a method of getting rid of the dreaded infinities. The procedure is known as 'renormalization'. Suppose someone is on a diet, but after a slap-up dinner find that they are one kilogram over the limit, and decide to hide it from their partner by adjusting the bathroom scale to read one kilogram under. Heinz Pagels, who was a professor at Rockefeller University and a well known science writer says, "This cheating – or rescaling is the renormalization procedure."[1]

By using the procedure the three theorists showed that the equations gave finite results that agreed with experiments to amazing precision. However, as Robert Adair an Emeritus Professor of Physics at Yale University observes, "the problem of the infinities was solved by sweeping them under the rug!"[2] The separate methods used by the three were then shown to be one and the same theory by the brilliant British physicist Freeman Dyson, whom we met before. He occupies the same post as Einstein did at the Institute for Advanced Studies in Princeton. Just as Dirac had done with Heisenberg and Schrödinger's theories, Dyson revealed a further secret power of mathematics, namely the power to describe the same underlying truth about Nature by quite different methods.

The theory became known as Quantum Electrodynamics or Q.E.D. for short. It provided a tiny scale quantum description of electromagnetism. In the theory, the electrons can be compared to two ice-skaters who are throwing a basketball between them. The ball they are throwing is the photon. As one electron catches the photon it is repelled away from its friend, showing the force of repulsion between them. To illustrate the interaction, Feynman came up with his space-time diagrams which became the standard way of describing all nuclear interactions.

Q.E.D. turned out to be the most accurate theory in the history of science so far, and still is. In 1965 Feynman, Schwinger and Tomanaga won the

Nobel Prize. The theory is accurate to within ten decimal places of results found in experiments. In 1985 Feynman described its accuracy like this: "If you were to measure the distance from Los Angeles to New York to this accuracy, it would be exact to the thickness of a human hair."[3]

Talking of Miracles

In their search for the Theory of Everything, physicists hoped there would be a way of showing that electromagnetism in the shape of Q.E.D. was actually one and the same force as the weak nuclear force. This seemed a pretty tall order. The weak force was about a billion times weaker than the electromagnetic force; and whereas the electromagnetic force (carried by photons) could operate like gravity to the ends of the universe, the weak nuclear force (beta decay) only operated at the minuscule atomic level.

In 1938 Yukawa (who discovered the pi-meson using maths), was working on the properties required for a messenger particle to carry the weak force when he came up with the W+ and W- particles. Unfortunately the theory was incomplete. However, in 1961 Sheldon Glashow at the University of California at Berkeley made a breakthrough with a third particle. It became known as the zed naught (Z^0).

The trouble with these particles was that, like the photon, they had no weight, yet experiment said they had to be quite massive by subatomic standards. Many people gave up, but an old high school classmate of Glashow's, Steven Weinberg (none other), then working at Harvard, and Abdus Salam, a Pakistani theorist working at Imperial College in London, continued independently despite the enormous odds.

To get over the problem of mass, they turned to the work of Peter Higgs at Edinburgh University, who had theorised the existence of what became known as the Higgs Field, and the particle associated with it as the Higgs boson. A property of this field was that it could give particles mass. Using the Higgs mechanism, Weinberg and Salam were actually able to unify the electromagnetic and weak nuclear forces. But nothing happened.

At first, their work fell on deaf ears. Although an enormous achievement, Weinberg's paper was only referred to by other scientists four times in the next four years. The reason was that their colleagues were sceptical, because no one had yet shown whether the theory would breed the dreaded infinities. It would mean checking each part of the theory by hand to see if all the infinities would actually cancel each other out. It would be a huge task.

In 1971, Gerhardt t'Hooft (Nobel prize in Physics 1999), a then twenty-five year old Dutch physicist at the University of Utrecht in Holland, had the bright idea of designing a computer program to search for the infinities in the theory. Using computers for this kind of thing was quite new in those days. From all the interwoven mathematical intricacy of the theory it seemed quite impossible that every one of the infinities would cancel each other out. Yet quite miraculously this happened!

From the mass of maths, the outcome of the programme was a perfect string of zeros. Every infinity cancelled exactly, proving beyond doubt that the theory was consistent. It was like a lightning bolt, proving how powerfully mathematics underlies the world.

New Machines to Prove the Maths

The cancelling of the infinities galvanised the physics world. At CERN, the Centre for European Nuclear Research where the Large Hadron Collider is now housed, they were able to predict what the masses of the W and Z particles were likely to be. They were big: eighty-seven times the mass of the proton for the W, and ninety-eight times for the Z^o. The reason they had never been seen before was because no accelerators yet built were powerful enough to reveal them.

New machines were then built, and old ones redesigned, in the hunt to try and find the W and Z. At CERN, Carlo Rubbia and a Dutch engineer, Simon Van de Meer, designed and built a proton-antiproton collider.

When they began in 1981, they had to create at least a billion collisions between protons and antiprotons to get results. They found a quick way of picking out 1 in 1 000 collisions that were useful. This left one million recorded collisions to examine more closely. Of these million collisions, just thirty-nine were selected, and of these just sixteen finalists were chosen. Of the sixteen, only five revealed the correct energy for a W particle. But they turned out to be just as Weinberg and Salam had predicted. Such is the power of mathematics.

In 1989 the Z° was discovered independently by two teams, one at CERN and another at SLAC, the Stanford Linear Accelerator, in California. Rubbia and Van der Meer shared the Nobel in 1984. Weinberg, Glashow and Salam had got theirs in 1979 before the actual confirmation of the Z° by experiment. This was highly unusual. The Nobel Committee is notoriously conservative and cautious before awarding prizes in science. Perhaps the power of maths to describe the physical world – so dramatically demonstrated by the infinities cancelling – had an impact on their judgement.

The discovery of the W and Z particles confirmed another great reduction in the number of forces in Nature. It united electromagnetism with the weak nuclear force, creating a single electroweak force. So now science was back to just three forces:

(i) Gravity
(ii) The electroweak force
(iii) The strong nuclear force

The Eightfold Way

For the next part of the story we have to go back to the early 1960's when experimentalists were building the first really powerful accelerators. By smashing known familiar particles like protons into each other at higher and higher speeds, they began to find more and more new particles never seen in Nature before, except perhaps at the Big Bang.

So many new ones were being discovered in the debris left over from these collisions that the whole neat picture of the atom was becoming thoroughly confused. Someone even suggested that, instead of winning prizes for new discoveries, people should be made to pay a fine! To the rescue came two theorists, Murray Gell-Mann from the U.S., and an Israeli, Yuval Ne'eman. Independently they found an eightfold pattern within the particles, each of which had different properties like mass, charge, spin, and a new property they called (not surprisingly), strangeness.

By creating this pattern, just like the Periodic Table of the elements in chemistry, it became possible to predict missing particles not yet discovered. At a physics conference held at CERN in 1962, Gell-Mann predicted the existence of a particle he called the Omega-Minus. Its properties, he said, would be:
  (i)   Same negative charge as the electron
  (ii) A mass of 3 288 times that of an electron
A strangeness of minus 3 units

In 1963 an accelerator at Brookhaven, New York, found the predicted particle which fitted the bill almost perfectly. Its mass worked out at 3 286 times that of the electron. It was an amazing confirmation of the theory, and yet further evidence of the power of maths. Gell-Mann called the new pattern the Eightfold Way, after the Buddhist Path to Enlightenment.

Just Three Basic Bits

After establishing that the eightfold pattern worked theoretically, the next step was to find out why it worked. In 1964 Gell-Mann and another U.S. physicist, George Zweig, independently suggested that the pattern could be explained if protons and neutrons were made up of even smaller objects. Zweig decided to call them 'aces', while Gell-Mann hit on the name 'quarks', after a phrase that stuck in his mind from James Joyce's novel 'Finnegans Wake', which went 'Three quarks for Muster Mark'. Quarks happened to win the day but aces would have done just as well. By assuming just three types of quark, they found that they could explain almost all the particles that had been found. For convenience, they called the different kinds of quark: 'up', 'down' and 'strange'.

For example, a proton is made up of one 'down' and two 'up' quarks, and a neutron of two 'down' and one 'up'.

At first, quarks were thought to be just helpful mathematical assumptions that happened to explain the eightfold way. The picture of the atom was so established that no one ever thought there could be an even smaller layer of reality hiding beneath protons and neutrons. It should be said that all the masses of new particles that spewed out of collisions in the big machines were extremely short lived, only surviving for millionths of a second before they decayed into the more familiar known particles, so they thought of quarks as a sort of purely mathematical device. The physicist and writer Richard Morris explains:

> Many physicists, including Gell-Mann himself, considered the quarks to be nothing more than useful mathematical fictions, not particles that had a real physical existence. In other words, the quark model was thought to be an abstract mathematical scheme, which made some predictions that could be confirmed by experiment, but which had no foundation in reality. [4]

Fooled Again

Another reason for reluctance was that, despite the thousands of collision experiments that had been carried out, no one had ever seen or smelled a trace of a quark. However, in 1968, by slamming high-speed electrons into protons at SLAC, they uncovered tiny point-like charges inside protons. Quarks turned out to be real. The power of maths to describe reality had notched up yet another glittering triumph.

Quarks had never been seen before because it required so much energy to break open a proton or a neutron, that new quarks were created in the process, through Einstein's equation $E = MC^2$. The quarks produced by this energy automatically turned into new protons and neutrons.

THE FINAL MYSTERY

It could now be seen that the strong nuclear force that binds protons and neutrons together in the nucleus, which Yukawa had first discovered, was actually the result of the force that held quarks together. When the quarks are no distance apart there is zero force between them and they sit snugly together. But as soon as you try to pull them apart the stronger the force between them becomes. And it's very strong.

Soon it was realised through experiment that there were more than just up and down quarks. They came in six different 'flavours' known as: up; down; strange; charm; bottom; and top. Just as electrons interact by exchange of photons, so quarks interact by exchange of messengers called 'gluons' – an appropriate name for a force that sticks the quarks so tightly together. Another way to imagine it is to go back to the comparison with ice skaters, except instead of them throwing and catching a basketball between them, they are now exchanging a heavy medicine ball – they have to skate much closer together because of the weight.

From accelerator experiments it had also been found that electrons belong to a family of particles known as leptons, or 'light ones'. The electron's brothers are the muon and the tau, and each of them is married to a neutrino partner, making six leptons in all. So it was now possible to say that everything in the universe is made up of just twelve bits or particles: the six quarks and the six leptons (and their antimatter opposites of course). However, we should also remember that the more exotic of the twelve - namely strange, charm, top and bottom quarks; and the leptons muon and tau are only ever seen in particle accelerator experiments. They don't occur naturally in the universe. The only time they did exist was at the moment of creation.

Every atom we encounter in Nature is made up of protons and neutrons with electrons zooming in orbit around them. So, it is fair to say that reality as we experience it is made up of just three bits of matter:

(i) Electrons
(ii) Up quarks ⎱
(iii) Down quarks ⎰ } protons + neutrons

Outline of the Standard Model

In the 1970s, theorists worked out how the force between quarks operated. With electromagnetism you have just the two familiar charges, positive + (plus) and negative – (minus); but with quarks the charges are more complicated, so they decided to give colours to the three kinds of charge. They named them after the primary colours, red, green and blue, which together make up the neutral colour white. Because of the colour charges the theory became known as Quantum Chromodynamics, or Q.C.D. for short. As mentioned earlier, the messenger particles which carry the colour force are known as gluons, and in the theory there are eight different kinds of gluon.

The theory is much more complicated than Q.E.D. and not nearly so accurate, but it brings us to a view of the world known as the Standard Model, which is considered to be reliable by most physicists. It does not yet explain gravity, which is thought to be carried by a messenger particle called a graviton, but despite many experimental tests gravitons have not yet been found.

The Standard Model goes like this:
1. All matter is made up of 12 elementary particles:
    (i) Six leptons
    (ii) Six quarks
2. All interactions between particles are carried by 3 forces:
    (i) Gravity – operates to the edge of the universe, carried by gravitons
    (ii) Electroweak Force – works in two forms:
        (a) electromagnetism – operates to the edge of the universe, carried by photons
        (b) weak nuclear force – operates at atomic level, carried by W+, W- and $Z^°$
    (iii) Strong Nuclear Force – operates at atomic level, carried by 8 gluons.

It is hoped that this picture can be simplified further by reducing all three known forces to one single force. Physicists think that at the extreme temperatures and energy levels that existed in the Big Bang, the electroweak force will merge with the strong nuclear force.

Theories that deal with this unification are known as G.U.T.s or Grand Unified Theories, but they still do not include gravity. It is hoped that by joining Einstein's General Relativity (gravity), with G.U.T.s (the other forces), we will finally find the Holy Grail, a theory of everything. Apart from the problem of gravity, the Standard Model relies heavily on the existence of the Higgs Field, named after Peter Higgs of Edinburgh University, whom we met before. In the Standard Model, it is the Higgs Field which gives all particles their mass or weight, and it should be manifested by a particle called the Higgs boson, which the press had dubbed the 'God particle'. Finding the Higgs was one of the main reasons for building the Large Hadron Collider. So it was to huge media acclaim that scientists at CERN confirmed the discovery of the Higgs boson on 14th March 2013. The Standard Model was finally complete. It proves that the mathematics discovered by Peter Higgs and others fifty years before, was a true description of our world long before we could prove it by experiment.

Back in the 1980's plans were drawn up to build the biggest ever particle accelerator. Located in Texas, the Superconducting Supercollider would have been the largest machine ever built, having a circumference of over fifty miles (eighty kilometres), and powerful enough to detect the Higgs particle. Unfortunately, because of the enormous cost, and to the dismay of the science world, the project was cancelled by the U.S. Government and work there was abandoned.

John Maddox, who was the editor of the famous science journal 'Nature' for many years, has said, "The importance of the Higgs particle cannot be exaggerated."[5] It must have been a long wait for Peter Higgs who shared the Nobel Prize in 2013. I once sat behind him at a concert, while in Edinburgh with my family, but never plucked up enough courage to speak to him, thinking that he may have been embarrassed by the unsolicited attention.

References

1. Pagels, Heinz., 1994. *The cosmic code.* London: Penguin. p 267.
2. Adair, Robert., 1987. *The great design.* Oxford: Oxford University Press. p 227.
3. Matthews, Robert., 1992. *Unravelling the mind of god.* London: Vintage Press. p 168.
4. Morris, Richard., 1992. *The edges of science.* London: Fourth Estate. p 14.
5. Maddox, John., 1998. *What remains to be discovered.* London: MacMillan. p 83.

Chapter Nineteen
Creation from Nothing

Where Things Meet

With the success of Q.E.D. and Q.C.D., theorists began to work on Grand Unified Theories or G.U.T.'s, which would link the strong nuclear force with the electroweak force. If they could show that these forces were somehow one and the same force, they would be able to reduce the number of forces in Nature to just two – gravity and the new strong and electroweak force.

"The basic idea of G.U.T.'s", Hawking explains, is that "the strong nuclear force gets weaker at high energies. On the other hand, the electromagnetic and weak forces...get stronger at high energies. At some very high energy, these three forces would all have the same strength and so could just be different aspects of a single force."[1]

The trouble is that the energy level at which grand unification takes place is enormous, so enormous that it would need a particle accelerator as big as the solar system. That, Hawking says, "would be unlikely to be funded in the present economic climate."[2] So physics was stuck with an apparently insoluble problem.

The resolution of this problem gave rise to the final union of physics with cosmology. It was realised that G.U.T. temperatures had only ever existed once before in the history of the universe – at the moment of its creation. One of the young theorists working on G.U.T.'s at Cornell University in the U.S. was Alan Guth. One day in 1978, when he had nothing better to do (he only had a passing interest in cosmology), he decided to drop in on a lecture being given by Robert Dicke of Princeton University. Dicke, you may remember, was the man who set out to find the microwave radiation from the Big Bang, but was beaten to it by Penzias and Wilson at Bell Laboratories, only a few miles away.

Dicke's lecture was about two of the main problems then facing cosmology. The first was the problem of critical density. We know that the universe is expanding as we sit here now, but what of the future? Will the universe go on expanding forever, or will the force of the Big Bang slowly weaken and eventually everything will stand still? And if it does, will there be enough matter in all the galaxies and clouds of gas for gravity to begin pulling everything together again, so that the universe begins to collapse back towards what cosmologists call the 'Big Crunch'?

In trying to work this out, they realised something quite amazing. If the universe has existed for 15 billion years already, then the critical density at the beginning must have been extraordinarily finely tuned to produce such a long lived universe in the first place – it would be like trying to balance a giant pyramid on its head. Stephen Hawking explains, "if the rate of expansion one second after the big bang had been smaller by even one part in a hundred thousand million million, the universe would have re-collapsed before it ever reached its present size."[3] Equally, if the rate of expansion had been greater by the same tiny amount, the universe would long ago have expanded out of sight. Most people would be inclined to say that this incredible accident that ultimately produced us, must point to the hand of God in the creation of the universe, but this is not the sort of magic solution accepted by science.

The second problem Dicke spoke about was known as the smoothness problem. As Penzias and Wilson discovered, and the COBE satellite later proved, the microwave background from the Big Bang is uniform in all directions. If you imagine standing anywhere on the Earth's surface on a clear night with your arms spread wide. Then, if you turn your head and sight along your right arm, out deep into the universe, and imagine looking 15 billion light years in that direction - then turn to your left, and imagine looking 15 billion light years in the other direction, you would actually be seeing a spread of 30 billion light years. But we know that the universe is only 15 billion years old. How then could the universe on our right hand know what the depths of the universe on our left was like? There wouldn't have been enough time in the history of the universe for one part to know what the other part was like, in order for it to be so similar everywhere, as revealed by COBE.

Alan Guth listened to Dicke with interest, and what he heard stuck in his mind over the next few years while he continued working on Grand Unified Theories. Dennis Overbye, a well known science writer, tells the story, "particle physics and cosmology had been moving closer and closer ever since Gamow had tried to use nuclear reactions to explain the big bang...The scientists who shuffled into a SLAC seminar room at the end of January in 1980 had little idea that they were about to witness the consummation of this curious slow marriage of ideas – the very large and the very small."[4]

Inflation

The main idea in particle physics was that, as the temperature increased, so the different forces in Nature melted into each other. At a certain temperature the electromagnetic force became the same as the nuclear weak force, and at an even higher temperature the electromagnetic and weak nuclear forces melted into the strong nuclear force, giving just one Grand Unified Force (not counting gravity).

These 'melting' processes are known as 'phase transitions', which are similar to the ones we are familiar with at home when ice melts into water, or water turns into steam. But one of the problems with G.U.T's was that they also predicted masses of things called magnetic monopoles (particles like the opposite ends of a magnet) which no one had ever seen. In thinking of a way to get rid of them, Guth thought about the possibility of another familiar occurrence called 'supercooling'.

If you cool water rapidly enough in the lab, it is actually possible to keep it liquid without it turning into ice even at twenty degrees below zero. The heat that should be given off by turning into ice, as happens with your refrigerator, is left hanging around as 'latent' or unused energy. The energy released by the freezing of a large swimming pool, for example, would actually be enough to heat a house for a couple of years.

Similarly, if the Higgs field had supercooled, there would be a large amount of energy left hanging around in suspended animation, or what is called a 'false vacuum'. When Guth plugged the supercooling bits into the normal equations for an expanding universe, what happened next must have made his hair stand on end. The universe exploded with a violence never before imagined in physics. Dennis Overbye describes it like this:

> With every tick of the cosmic clock (a tick in this case being 10^{-34} seconds) the size of the universe doubled...In no time at all – a millionth of a trillionth of a trillionth of a second – the universe would double in size 100 times, becoming a trillion trillion times larger. Guth had a runaway universe on his hands. Disrupt the big bang? Letting the Higgs field supercool into a false vacuum state was like setting off an atomic bomb in the midst of a hand grenade explosion.[5]

This previously unimagined ferocity was later called 'Inflation' by Guth. What it did was to wipe out any possibility of masses of monopoles – but at the same time, it also cured cosmology's twin problems of the incredible fine tuning of the critical density (the pyramid standing on its head), and the smoothness problem. As Weinberg explains, "without such a supercooling era, it would be very difficult to understand why the microwave radiation background from points in opposite directions in the sky has the same temperature...without a supercooling era there would not have been time in the history of the universe for any influence to have reached these points from any common source."[6]

The Ultimate Free Lunch

The first physicist to think about the possibility of the universe being created out of nothing at all suffered from what must be a difficult handicap for any scientist, let alone one proposing such an astounding theory. His name was Edward Tryon. He gained his doctorate under Weinberg and, as an assistant Professor at Columbia University in 1969, he went to hear a British cosmologist, Dennis Sciama, giving a lecture.

THE FINAL MYSTERY

As he was listening, his thoughts began to wander. He was thinking about the quantum vacuum which, as we discovered, is actually a seething mass of virtual particles popping in and out of existence – first measured by Casimir. "Suddenly he was seized by an idea, and was startled to hear himself interrupting Sciama's talk. 'Maybe' he blurted out, 'the universe is a vacuum fluctuation.' Tryon's colleagues laughed...' It just cracked them up', Tryon recalled...' I was deeply embarrassed...I never told them I'd not been joking.'[7]

Three years later, while sitting quietly at home one evening, he realised how it might actually work. It was a revelation, he later said, that made a chill run through his body. In 1973 he published a paper in 'Nature' entitled 'Is the Universe a Vacuum Fluctuation?' He had worked out that the universe might have zero energy overall. If it had zero energy at the very beginning, then a hiccup in the vacuum might suddenly grow from a tiny quantum event into a huge universe. Basically, all the positive energy locked inside the stars, gas and galaxies, equals the negative energy represented by gravity. Again, I find Hawking's explanation the easiest to understand: "In the case of a universe that is approximately uniform in space, one can show that this negative gravitational energy exactly cancels the positive energy represented by the matter. So the total energy of the universe is zero."[8]

This means that no physical laws, like the law of the conservation of energy, are violated; and it also means that once (and only once) in the history of the universe it was possible to get something for nothing. Creation from a quantum fluctuation of the vacuum provides the platform for Guth's inflation. About the inflationary model, he says that it offers, "the first plausible scientific explanation for the creation of essentially all the matter and energy in the observable universe." Although I imagine he regrets it now, he is famous for saying, "that the universe may be the ultimate free lunch."[9]

Paul Davies puts it neatly: "The thorny problem of what caused the big bang is therefore solved by the inflationary theory: empty space itself exploded under the repulsive power of the quantum vacuum."[10] As I said earlier, when I was growing up, all ideas about how everything began and where the universe came from belonged entirely to religion and theology. It is truly astonishing then that science has been able to come up with a logical and consistent explanation. As Davies later says, "the idea of creation from nothing has, until recently, belonged solely to the province of religion."[11]

However, as I said at the start, there is one slightly horrendous complication. If inflation is correct, it also inevitably means that our whole observable universe must be but an extremely diminutive dot of a much bigger expanse called a 'multiverse.' Dennis Overbye explains that because of inflation our universe is:

only a speck in a bigger, wilder ensemble… Our own inflated bubble was probably trillions, not billions, of light-years across. And beyond that could be other bubbles, other enormously inflated island universes undergoing their own sagas of expansion and evolution…in their own separate space-times.[12]

About inflation, John Maddox says, "The philosophical cost of this innovation, however, has been huge and excessive…inflation inescapably requires that there must be a multitude of other universes alongside our own, each derived from a different speck of space-time and each evolving in parallel."[13] Actually a mere 'multitude' is being very conservative. We will see in the next chapter that there are almost certainly an infinite number.

At this point I'd like to introduce you to an all time hero of mine. I have actually corresponded with him in the distant past. Talking about heavyweight champions of science, they don't come with more gravitas than Lord Rees of Ludlow, better known as Sir Martin Rees, first knighted as Britain's Astronomer Royal and now made a peer, who was not only Master of the temple of Trinity College Cambridge, but also President of the Royal Society. He appears frequently on science documentaries. So it's worth listening to what he says. He sums up this gigantic leap in human consciousness in three sentences. Here it is. He says that our universe, "may stretch not just millions of times further than our currently observable domain, but *millions of powers of ten* further. [His emphasis] And even that isn't all. Our universe, extending immensely far beyond our present horizon, may itself be just one member of a possibly infinite ensemble."[14]

It's a quote I love. It robs me of breath each time I think about it.

References

1. Hawking, Stephen., 1988. *A brief history of time.* London: Bantam Press. p 74.
2. ibid., p 74.
3. ibid., p 121.
4. Overbye, Dennis., 1993. *Lonely hearts of the cosmos.* London: Picador. p 247.
6. ibid., p 242.
7. Weinberg, Steven., 1983. *The first three minutes.* London: Flamingo Fontana. p 159.
8. Ferris, Timothy., 1990. *Coming of age of the milky way.* London: Vintage. p 354.
9. Op. Cit., Hawking, p 129.
10. Guth, Alan., and Steinhardt, Paul. 1989 *The inflationary universe.* The New Physics ed. by Davies, Paul., Cambridge: Cambridge University Press. p 54.
11. Davies, Paul., 1985. *Superforce.* London: Unwin. p 193.
12. ibid., p 195.
13. Op. Cit., Overbye, p 257.
14. Op. Cit., Maddox, p 54.
15. Rees, Martin., 1999. *Just six numbers.* London: Weidenfeld & Nicholson. p 11.

Chapter Twenty
The Multiverse

Dark Matter

We have to go back a bit for the next part of the story. Back to 1933 to a Swiss-American with the prickly sounding name of Fritz Zwicky. He was a professor at the world famous California Institute of Technology. By all accounts he was a bit of a character. Although he lived in the U.S.A. for over thirty years he never gave up his Swiss citizenship. He didn't get on too well with a few of his colleagues, and when he didn't like someone he would refer to them as 'spherical bastards'. When asked why spherical? His reply was that they were bastards from every angle!

Together with Walter Baade he was responsible for the installation of a new telescope at Mount Palomar which could map large areas of the sky in one go. With this Zwicky was able to carry out the first big survey of thousands of galaxies. It was Zwicky who discovered that galaxies form clusters, and clusters form super clusters. But his observations revealed something strange. From the way the galaxies were moving, there had to be a lot more matter in them than could actually be seen or they would fly apart and not form groups. He called it 'dark matter' because it did not give off any light. He published this work but no one paid much attention to it. Zwicky was the first to coin the term 'supernova' and guessed that they were exploding stars. Apart from his astronomy he was famous for the development of jet and rocket engines after the Second World War.

The next character to enter the picture is Vera Rubin. She worked for the Carnegie Institute and her speciality was measuring the light signatures from different galaxies. Over many years collecting evidence she noticed something that definitely should not be happening. According to the rules of gravity the stars at the outer edges of galaxies should be moving more slowly as the galaxy revolved, just like the outer planets orbit the

THE FINAL MYSTERY

sun more slowly than the inner planets. But they weren't. Vera discovered instead that they were moving just as fast as the stars nearer the centre and should therefore be flying off into space. Something was badly wrong.

She presented her research to the American Astronomical Society in 1975. Unfortunately, the subjects of both her masters' degree and her doctoral thesis had been on controversial issues and people were cautious about accepting her work. This is not surprising really, because it meant that either Newton's laws of gravity did not apply on a galactic scale, which would be like throwing a huge spanner in the works, or there was some mysterious matter holding the galaxies together. She worked out that for the galaxies to be behaving as they were, it meant there had to be far more invisible stuff than the luminous matter we could see. This was a lot for people to accept. One wonders if there wasn't also a bit of male chauvinism going on. But then people remembered Zwicky's work back in the 1930's.

Over the years more and more observations confirmed Rubin's findings and astronomers had to come to terms with it. Zwicky's name for it has stuck and it is now called 'dark matter'. Although it can't be seen, astronomers can tell where it is from its gravitational attraction. It seems that each galaxy is surrounded by a large halo of dark matter much bigger than the galaxy itself.

So what is dark matter made of? Well, nobody knows for sure, which is pretty exciting, but they have managed to come up with two comical candidates known as WIMP's and MACHO's. WIMP's stands for 'weakly interacting massive particle' and MACHO's stands for 'massive compact halo object'. It is thought that MACHO's are made up of several different kinds of objects like brown dwarf stars (stars that have failed to ignite), black holes, planets, asteroids, clouds of non luminous gas and mini galaxies that are too dim to be seen. The WIMP's are particles much smaller than atoms. A big possibility is that dark matter is made up of neutrinos, which we saw can pass straight through a planet, and so are very difficult to detect. There are also other exotic particles called axions. Astronomers and physicists are searching for them deep underground which might seem a little odd, but they have to filter out all the normal cosmic rays. Most of them will be stopped by the solid rock, so what is left could be the particles of dark matter.

It turns out that 90% of all matter in our universe is this dark matter, and the stuff we can see makes up only 10%. Quite a remarkable situation. But cosmologists were about to discover that matter made out of atoms was only a small part of the whole story.

Dark Energy

Ever since Hubble had discovered that the universe was expanding, people thought that the expansion of the universe must be slowing down after the initial explosion of the Big Bang. In the 1990's two teams of cosmologists and physicists decided to try and measure the speed at which the universe was actually slowing down. To do this they used supernova in very distant galaxies to make their measurements, so the work was called The Supernova Cosmology Project.

The first team was situated at the Lawrence Berkeley National Laboratory in California and the second team was in Australia. After nearly ten years of collecting data, they announced their results in 1998. What they had discovered shocked everyone. Instead of slowing down, the expansion of the universe was speeding up! Some extremely powerful unseen force was working against the pull of gravity. It was like something straight out of the log of 'The Star Ship Enterprise'. Except it wasn't science fiction. It was real and it involved the whole universe. They called it Dark Energy.

You'll remember how Einstein added an extra bit to the equations of General Relativity to keep the universe still because he didn't like the idea of it expanding? He called it the 'cosmological constant', and now it looks as though he might have been right all along by pure accident. Dark Energy is actually the energy of space itself. This is where the quantum vacuum comes in again. It consists of 'virtual' particles jumping in and out of existence. Although they are virtual, they still carry energy as we saw from the Casimir Effect. It acts exactly like the bit that Einstein added to his equations, so the vacuum energy is now called the cosmological constant. This is what 'dark energy' is thought to be, and it is causing the expansion of the universe to speed up.

Since the Supernova Cosmology Project there has been further independent confirmation of their results by more accurate measurements of the supernova, and by a technique called gravitational lensing. Also, four different groups of cosmologists found that the gravitational effect of dark energy has actually slowed down the collapse of certain rather dense groups of galaxies. Then in June 2001 NASA launched another telescope that was much more sensitive than COBE to analyse the cosmic microwave background in more detail. It is called WMAP, short for the Wilkinson Microwave Anisotropy Probe. Just as COBE had done, it mapped the sky for the heat left over from the Big Bang. It was able to measure temperature variations to 1 part in 100 000 showing that the universe is the same everywhere you look, and that these slight ripples and bumps are consistent with the seeds of cosmic structure that later formed the galaxies, stars and planets.

One of its biggest discoveries was to confirm fairly convincingly that the theory of inflation is correct. But it also finally established what the universe is made of:

4% Atoms: The stuff we can see, galaxies stars, gas and, lately, planets.
22% Dark matter: WIMP's and MACHO's.
74% Dark energy: The energy of empty space.

Lord Rees, the Astronomer Royal whom I mentioned earlier, came up with this cracker in a BBC Live Chat Special. He pointed out that, "the atoms we are made of are, as it were, just a strange contaminant of the universe whose dynamics and gravity are dominated by quite different stuff."[1]

A Very Special Universe

Before cosmologists discovered that the expansion of the universe was accelerating, theoretical physicists had tried to work out the energy within empty space. It turned out to be infinite which didn't help, especially as they thought it was close to one. So they made an educated guess. The result was almost as bad. Steven Weinberg described it like this, "the vacuum energy per volume comes out to be enormously large: it is about a trillion trillion trillion trillion trillion trillion trillion trillion trillion trillion times larger than is allowed by the observed rate of expansion. This must be the worst failure of an order-of-magnitude estimate in the history of science."[2]

It was naturally a serious embarrassment, but most cosmologists and physicists thought that, as often happens, something would be found to make all the positives and negatives cancel out (as they often do), and they would be left with a precisely zero vacuum energy. So all they had to worry about really was gravity. Unfortunately for everyone it has now been found that things don't quite cancel out to zero. Instead they cancel out to a phenomenal 10^{120} decimal places. This is more than an extraordinary event, it is scarcely believable. It is an inconceivably tiny amount.

Leonard Susskind, the father of string theory who we will also be hearing about shortly says, "as absurd as it seems, the vacuum energy exactly cancels for the first 119 decimal places but then in the 120th place, bingo! A bit of vacuum energy. How can such a situation possibly be explained?"[3] It means that our universe is far, far more improbable and therefore special than anyone could ever have imagined.

Ever since the 1960's scientists had realised that had the basic structure of our universe been only slightly different from what it is, then it would not have been possible for life to have evolved. The Standard Model has over twenty things that scientists can't account for. These are things like the speed of light, the strength of gravity, the exact charge on the electron, the mass of the proton and why it is precisely 1 800 times 'heavier' than the electron, and why the forces like the strong and weak nuclear forces have the particular strengths that they do.

The whole purpose of a Theory of Everything is to explain why these 'constants of nature' are the way they are. But the amazing thing is, if any of them had been any different from what they are, then we wouldn't be here. If gravity had been stronger than it is, then all the stars would have burnt out their fuel much too quickly for intelligent life to emerge. If it had been any weaker then galaxies and stars could not have formed. If the strong nuclear force had been any different, then all the hydrogen would have converted to helium before anything could get started.

This line of thinking is called the Anthropic Principle, because it seems that the universe was very precisely designed to create intelligent life. There are literally hundreds of examples of this design. Take ordinary water, H_2O. This simple combination of two hydrogen atoms and one oxygen atom has a unique property. Unique in the universe. With no other substance is the solid form lighter than the liquid form. With everything other than water the solid state is always heavier than the liquid state. But if ice was not lighter than water because it expands, then the oceans would have frozen solid long ago and life would never have started.

Take carbon, a seriously important element as far as our bodies are concerned. The way it is formed inside stars is unlike the formation of any other element. It requires three helium nuclei to come together at exactly the same instant in a complicated dance that wouldn't be possible if the forces involved had been even minutely different.

Going back to the astonishingly precise value of the vacuum energy, Paul Davies comments like this. "The cliché that 'life is balanced on a knife-edge' is a staggering understatement in this case: no knife in the universe could have an edge *that* fine…it would be an extraordinary coincidence that *that* level of cancellation – 119 powers of ten, after all – just happened by chance to be what is needed to bring about a universe fit for life……….That level of flukiness seems too much to swallow."[4]

The defining book on the subject was published by John Barrow and Frank Tipler in 1986. The introduction was written by John Wheeler, whom we met before whose students included Richard Feynman and Hugh Everett. This is how he described the Anthropic Principle. "It is not only that man is adapted to the universe. The universe is adapted to man. Imagine a universe in which one or another of the fundamental dimensionless constants of physics is altered by a few percent one way or the other? Man could never come into being in such a universe. That is the central point of the anthropic principle. According to this principle, a life-giving factor lies at the centre of the whole machinery and design of the world."[5]

Here is another quote that I'm fond of. It's from no lesser person than Freeman Dyson. He says, "As we look out into the universe and identify the many accidents of physics and astronomy that have worked together for our benefit, it almost seems as if the universe must in some sense have known we were coming."[6] This is what I mean when I talk about the enormous responsibility of being human.

You may well be saying to yourself, "so there *must* be an Ultimate Designer then?" It seems very much that way. So much so that I can't understand why those who follow Creationism and Intelligent Design don't use this as their ultimate starting position. Most especially in the case of the vacuum energy.

Many Universes

But, as I said before, this is not the way of science. One of the side effects of Inflation, as we saw earlier, is that it predicts an infinity of other universes. Since Alan Guth discovered it, more and more evidence has been gathered in support of the theory, so that most of the major figures in cosmology and physics now believe that inflation is correct, and along with it, that there are an infinite number of universes out there. They call it 'The Multiverse', and casually discuss it over their cups of coffee!

We have certainly come a long way. From the limited horizons and consciousness of our hunter gatherer ancestors who thought the world was flat, to the discovery that our Earth is round and revolves. And then, that it is only one of eight other planets orbiting an ordinary star in the outer suburbs of a single galaxy, in a universe containing hundreds of billions of galaxies. And now, that our universe is but one of an infinite number. Fortunately this is as large as the mystery of our existence can possibly get. You can't get anything that is bigger than an infinity. That's why we have been 'sizing them up' so to speak, as we've been making our way.

Although it is possible to discuss it over a coffee or a glass of wine, or jar of beer, when you actually undertake the mental exercise of imagining that we really *do* live within a multiverse, it becomes utterly astonishing. I mean, as astonishing as anything can get. In fact, it is not possible for there to be anything greater or more awe inspiring, than an infinite number of universes.

One of the first things the multiverse does is that it explains the Anthropic Principle. Remember that if the speed of light or the strength of gravity had been any different from what they are then we wouldn't be here. And more especially, if the energy of empty space hadn't been so incredibly fine-tuned, we wouldn't be here either. This makes a Grand Designer a serious contender. But if there are an infinite number of universes then it's not so surprising that one of them should suit us. It's not so surprising therefore that one of them has exactly the right combination of laws and 'constants' to evolve from the first atoms of hydrogen to the human brain. Lee Smolin, one of world's leading physicists, explains: "So there is a simple answer to the anthropic question. Among all the possible universes, a minority will have the property that their laws are hospitable to life. Since we are alive, we naturally find ourselves in one of them."[7]

So it looks very much as though it requires an infinite number of universes just to produce us. For me that screams a big question. How much more important does that make our tiny smudge of consciousness? I suggest that it makes it vastly more important than anyone has ever considered previously.

Sir Martin Rees has a simple and clear analogy to describe why we find ourselves here. "The cosmos may have something in common with an off-the-rack clothes shop: if the shop has a large stock, we are not surprised to find one suit that fits. Likewise, if our universe is selected from a multiverse, its seemingly designed or fine-tuned features would not be surprising."[8] What is seriously profound, is the realisation that if it requires an infinite number of universes to produce us – then how much greater must be our destiny? I can't help wondering what the Patriarchs of the Old Testament would have made of this.

Different kinds of Multiverse

For scientists there are several different kinds of multiverse. In some, the laws of Nature are very much the same as ours, only slightly different. In others the constants like the speed of light and the strength of gravity are very different, which produces physically different kinds of universes.

Some cosmologists believe that our own universe is itself infinite, made up of an infinite number of regions called Hubble Volumes. A Hubble Volume is quite fun, it goes like this. The light from the furthest objects we can see right now are roughly 14 billion light years away. But because the universe is expanding, what we see tonight is by now (by the time the light reaches us) actually another 14 billion light years further away, making them currently about twenty-eight billion light years away. So those most distant galaxies are actually at this point way beyond our horizon. Also, because the speed of the expansion is getting faster, those most distant galaxies are calculated to be some 40 billion light years away from us in a big circle in every direction away from the Earth. This circle is one Hubble Volume.

The cosmologists I'm talking about think that there are an infinite number of these Hubble Volumes. And if there are an infinite number then, strange as it may seem, there *must* be – very far away, can you believe, an exact copy not just of the Earth but also of every single one of us. The nearest copy of you is said to be some $10^{10,29}$ metres away. Why they choose to measure it in metres I have no idea. But it doesn't stop there either. Because we are talking about an 'infinity' here, there is not just one identical copy of yourself and myself out there, there are actually an infinite number of copies of each of us! That's pretty astonishing. But it gives us an idea of just how powerful an infinity is. John Barrow, the professor of Mathematical Sciences at the University

THE FINAL MYSTERY

of Cambridge, describes it like this. "Thus at any instant of time – for example, the present moment – there must be an infinite number of identical copies of each of us doing precisely what each of us is now doing. There are also infinite numbers of identical copies of each one of us doing something other than what we are doing at this moment. Indeed, an infinite number of copies of each of us could be found at this moment doing anything that it was possible for us to do with a non-zero probability at this moment."[9] That's how powerful an infinity is.

You may feel the need to take a deep breath here. I want to make a couple of comments. The first is just, well, that's what happens when you mess with infinities. They are *big*. Fantastically *big*. The other point is to remind you yet again (to the point of not being too irritating I hope), that this is serious science. And science is the only tool we possess for seeing the world. Without it we would still be convinced by the evidence of our eyes that the Sun goes round the Earth.

And here's another interesting thought. People love to speculate that there are many other extraterrestrial civilisations out there on other planets within our own universe. And also, that some of them may be vastly more advanced than our own. But no matter how advanced they may be technologically, they can never progress beyond – they can only reach as far as, things that are infinite. It's an ultimate limiting factor, like the speed of light – and we are already quite familiar with things infinite! Even if their civilisations are more sophisticated, and let's hope that they are, then considering our knowledge of the infinite, their consciousness can only make them our brothers/sisters-in-arms.

THE MULTIVERSE

Figure 11 Illustration of chaotic inflation

The most popular kind of multiverse comes, as I said, from Inflation. It pictures the multiverse arising from the foam of the vacuum in an ever increasing number of bubbles, and each bubble becomes a different universe. In some the laws are very similar to our own but only slightly different, in others they are wildly different, even with different dimensions. In some, matter and antimatter exactly cancel each other out and: **Bop!!** ... nothing happens. In others gravity is much weaker and they expand out of sight. In still others it is much stronger and they re-collapse abruptly. But each one has its own space and time. This produces quite a queasy feeling in your stomach, because it means that they are all happening at the same 'moment' so to speak, but outside our own space and time. So they must be happening right now all around us, perhaps a millimetre from our faces, but we wouldn't be aware of them because they are not within our time and space. This suggests that space and time are somehow something 'secondary', a kind of product of something else. Can maths be far away?

The bubble multiverse idea is known as chaotic or eternal inflation and came first from the Russian born cosmologist Andrei Linde at Stanford University in the U.S. Here's how Max Tegmark, a Swedish American cosmologist from the famous MIT, describes it. "Each such bubble is infinite in size, yet there are infinitely many bubbles since the chain reaction never ends. Indeed, if this exponential growth of the number of bubbles has been going on forever, there will be an uncountable infinity of such parallel universes."[10]

The other main kind of multiverse goes back to Hugh Everett and the 'Many Worlds' interpretation of quantum mechanics which we discussed in chapter sixteen about the Great Smoky Dragon. In this version all the laws are the same as our universe, but everything within them exists in an infinite number of states. This is why the 'superposition' doesn't collapse when we measure something. There will be more about this in chapter twenty-one when we look at quantum computers.

String Theory

There is a whole other angle to the multiverse and it comes from String Theory which as I said earlier, is the most promising candidate we have for a Theory of Everything. It's the theory that's able to link the big world of gravity with the small world of quantum particles. It was first discovered by Leonard Susskind and Yoishiro Nambu back in 1969 and created great excitement. If you recall, Weinberg described it as having the smell of inevitability. Everyone thought *this* was going to be it. A final theory was in sight. In his opening speech, when he was made Lucasian Professor at Cambridge in 1979, Hawking predicted a final theory quite soon.

Unfortunately it didn't work out that way. However, it is still the 'chosen' route to find the Holy Grail. Someone estimated that it still attracts nine out of ten theoretical physicists to its cause, but over time it has thrown up increasing problems. The latest versions are called M-Theory. The M is said in some quarters to stand for 'mystery'! The main problem is that it describes millions and millions, (actually trillions and trillions) of different universes, each with its own constants and basic laws, and there is no way to tell which one is ours.

Susskind says that when there were just a million or so solutions he wasn't particularly worried, but the tipping point came for him (and I guess many others), when in the year 2000 two theoretical physicists, Joe Polchinski and Raphael Busso, published a paper showing that there might be somewhere in the region of 10^{500} solutions! That's when he started to have serious doubts! You can imagine why.

This is an absolutely colossal number. Bear in mind that atoms are pretty tiny objects, and then think about all the masses of stars and galaxies we can see in the universe stretching far out. Well, in all that matter there are only 10^{80} atoms. If you think back to how 'powers of ten' actually work, then you'll see that 10^{500} is a horrifyingly monumental number. Not for mathematics of course, it can handle any number. I wonder why?

You have to ask yourself whether String Theory isn't describing every possible – every mathematically conceivable universe… which means an infinite number of them. Back in the 1980's people thought that String Theory was so exciting they described it as something from the 21st century that had accidentally fallen into the 20th by mistake. I can't help thinking that String Theory might be correct and that it is simply describing the many faces of the multiverse.

The last word for now on the multiverse should come from Steven Weinberg who is regarded by many as the patriarch of modern theoretical physics. It's a friendly anecdote, and comes from his opening talk entitled 'Living in the Multiverse', which he gave to a symposium on 'Expectations of a Final Theory' which was held at the temple of Trinity College, Cambridge, on 2nd September 2005. He concludes his talk by relating how at the airport back home in Texas on his way to the meeting, he noticed an article in a magazine. I'll let him tell it. "Inside I found a report about a discussion at a conference at Stanford, at which Martin Rees said that he was sufficiently confident about the multiverse to bet his dog's life on it, while Andrei Linde said he would bet his own life. As for me, I have just about enough confidence about the multiverse to bet the lives of both Andrei Linde *and* Martin Rees' dog."[11]

References

1. Whitehouse, David., 24th Jan. 2002. *Sir Martin talks multiverses.* BBC: Live Chat Special.
2. Weinberg, Steven., 1993. *Dreams of a final theory.* London: Vintage. p 179.
3. Susskind, Leonard. 1 Nov. 2003. *A universe like no other.* New Scientist No. 2419. p 34.
4. Davies, Paul. 2006. *The goldilocks enigma.* London: Allen Lane. p 170.
5. Barrow, John., and Tipler, Frank. 1986. *The anthropic cosmological principle.* Oxford: Oxford University Press. p vii.
6. Dyson, Freeman. 1979. *Disturbing the universe.* New York: Harper & Row. p 150.
7. Smolin, Lee. 2000. *Three roads to quantum gravity.* London: Weidenfeld & Nicholson. p 199.
8. Rees, Martin. 2001. *Our cosmic habitat.* London: Weidenfield & Nicholson. p 165.
9. Barrow, John. 2005. *The infinite book.* London: Vintage. p 156.
10. Tegmark, Max. 2004. *Parallel universes.* In: Barrow J, Davies P, Harper Jr. C, Editors. *Science and ultimate reality.* Cambridge: Cambridge University Press. p 466.
11. Weinberg, Steven., September 2 2005. *Living in the multiverse.* From symposium, EXPECTATIONS OF A FINAL THEORY. Trinity College, Cambridge. ArXiv:hep-th/0511037v1 3 November 2005.

Part Five

The Biggest Mysteries

Chapter Twenty-One
The Mystery of Parallel Universes

Taking Stock

In Part One we saw how everything in the universe is just the result of evolution towards gradually more complex arrangements of matter. It started with the vacuum and the first simplest atoms of hydrogen and helium, which led to the first galaxies and then to the death of the first generation of stars. In the process of burning and dying they formed all the heavier, more complicated, atoms including the carbon out of which we are made.

We followed how the earth formed and how different kinds of atoms began to link arms to make the still more complicated structure of molecules that led to the evolution of the first primitive life forms, and eventually to the most complex arrangement of matter known to us in the universe, the human brain. We considered the first revolutions like the use of fire, the domestication of crops and animals, the growth of cities into states, the invention of writing and mathematics, and technological innovations like bronze and iron and the invention of the wheel.

In Part Two we traced the story of astronomy from the classical Greeks to the present day. And we saw how each new revolution in knowledge enlarged our horizons by leaps and bounds, from thinking the Earth was the centre of the universe to discovering that our sun is just one star in the Milky Way, and that there are billions of other galaxies besides our own. We covered the discovery that the universe is actually expanding and that science is now able to measure the leftovers of the very beginning, as well as map the giant clusters of galaxies.

In Part Three we examined the other end of the scale which took us to the world of the atom and beyond. We covered the major discoveries about the atom and how this led to the Quantum Revolution and how its meaning has been interpreted in different ways.

In Part Four we followed the discoveries of particle physics and the development of the Standard Model and how these discoveries helped to understand the way the universe evolved. This demonstrated that everything from the incredible smallness of the Planck Scale to the outer edges of the universe, fourteen billion light years in all directions, is actually a single joined up entity, and that it all came from the tiniest hiccup in the vacuum.

Strange Encounters

Along the way there have been some strange encounters, but in Part Five we are going to find that our reality is even stranger still. Back in chapter fifteen we examined the different ways of trying to understand the quantum world. The central problem, bizarre in itself, was that when bits of atoms travel they go by every possible path or an infinite number which is known as a superposition. But as soon as we measure where the bit is, the wave of all possible routes collapses down to a particular one. This is the mystery of the 'Great Smoky Dragon'.

Very briefly, the official interpretation maintained that you could not say anything about the dragon. The next came from Eugene Wigner who made human consciousness responsible for the collapse of the wave function, so that we create reality when we observe it. Famously, Einstein said he couldn't believe that the moon did not exist when he wasn't looking at it. The world is so strange that this view has not been abandoned.

Then there was the suggestion that there might be a set of hidden variables that might pilot the particles from one place to the next. This worked, but it had what seemed to be a big disadvantage because it meant that every particle in the universe had to be connected. Last there was the 'Multiple Universe' or 'Many Worlds' interpretation which said that there was no problem, because every possible path was actually real. Each time an observer measures an atomic object, both the observer and the object split into different parallel universes. In one of them Schrödinger's cat is alive, and in another it is dead. Every possible path does actually exist, which means that every time one bit of an atom collides with another, which is happening countless trillions of times a second everywhere, the universe is splitting into trillions of parallel universes. In some of them Timbuktu wins the world cup, and in others Elvis is still alive.

The Believers

This is weird. I mean, sensationally weird. And yet it is the one which most physicists and cosmologists believe to be real. This is probably because it is the most 'straight forward'. It doesn't require the superposition to collapse at some unexplained moment created by the act of 'observing'. Every now and then someone takes a poll at international conferences to test this. One of the first was taken, as I said before, way back in the 1980's by David Raub, a political scientist. At an important gathering he asked seventy-two scientists whether they believed in the Many Worlds interpretation and found the following:

1. 58% thought it was true.
2. 18% did not think it was true.
3. 13% thought it could be true but were not convinced.
4. 11% had no opinion.

Among the 58% who thought it was true were four Nobel Prize winners, including Steven Weinberg, and world renowned cosmologists like Stephen Hawking. In his latest book for the general public Hawking actually says, 'this idea that the universe has multiple histories may sound like science fiction, but it is now accepted as science fact.'[1]

If this is not confirmation enough about the multiple parallel worlds that surround us, then the new science of Quantum Computing should finally convince even the most sceptical.

The Birth of Quantum Computing

The story begins in 1982, when Richard Feynman produced an abstract model of how a quantum system could be used to perform calculations. Nobody thought much of it at the time because he had come up with several ideas that seemed crazy at the time, like the fact that antimatter is actually ordinary matter moving backwards in time. Then in 1985 David Deutsch, of Oxford University, came up with a ground breaking theoretical paper describing how any physical process could be modelled perfectly using a quantum computing system.

The next big step was taken in 1994 by Peter Shor at Bell laboratories (remember Penzias and Wilson?). He was able to show how a quantum computing machine would be able to factorise any number, no matter how big.

This immediately caused a bit of a stir. The reason is that all the secret codes used by the government and military rely on numbers that are impossible to factorise in practice. What they do is to get two huge numbers of say, 150 digits in each number, and multiply them together to get an enormous number which they then use to send the message. To read the message the enemy has to know the two numbers they had chosen in the first place. In principle it would be possible to do by painstakingly examining every possible combination of numbers, but with such huge numbers this would take even the fastest digital computer longer than the age of the universe to work out.

How They Work

Not so with a quantum computer, and hence the interest of the military who don't want their codes broken. Basically, quantum computers do calculations using atoms or atomic particles instead of computer chips. Atoms have a natural spin that can be either 'up' or 'down', which coincides neatly with the current digital binary method which represents everything by a series of '0's and '1's. With an atom, a spin pointing up can represent a '1' and pointing down can represent a '0'. Changing the spin 'up' or 'down' is the same as flipping the switch on and off on a tiny transistor.

The difference, as we have seen, is that atoms exist in a superposition of an infinite number of states at one and the same time, each within a separate parallel universe. So a quantum bit or 'qubit' as it's known, can be either a '1' or a '0' or a superposition of both simultaneously. This means that instead of solving a problem by adding all the numbers in sequence, a quantum computer can add all the numbers at once. Deutsch explains that quantum computing, "will be the first technology that allows useful tasks to be performed in collaboration between parallel universes. A quantum computer would be capable of distributing components of a complex task among vast numbers of parallel universes, and sharing the results."[2]

The next step forward came in 1996 when Lov Grover, also at the renown Bell labs, came up with what is known as Grover's Search Algorithm, which showed how a quantum computer could search a database millions of times faster than a conventional digital computer. It meant that quantum computing hit the big time. It was not just restricted to cracking obscure codes, it could be widely useful in obtaining information quickly from huge databases, a common everyday task of most conventional computers.

What They Can Do

For example, if a database contained say, a million items. Then it would take a conventional computer five hundred thousand steps to find the right item, whereas a quantum computer would need only a thousand steps. Another example that is often given is the old one of having to find a needle in a haystack. A conventional computer would have to examine each strand of hay individually and in order, whilst a quantum computer could identify every strand of hay at once.

Isaac Chuang, a professor of physics at the famous Massachusetts Institute of Technology, is one of the most successful experimenters working in the field. The example he gives is this. If the most powerful supercomputer in existence today had to find a single telephone number in a database consisting of all the phone directories in the whole world, it would take a month. A quantum computer could do it in twenty minutes.

As one of the pioneers himself, Deutsch says that the first working quantum computer was actually built way back in 1989. It was an extremely basic device, but after the discoveries of Shor and Grover, an international race was on to build the first working commercial quantum computer.

Where It's Happening

There are many university laboratories all over the world as well as big companies like IBM and Hewlett-Packard engaged in the race. There are even collaborations between universities and public companies. Isaac Chuang leads a team of scientists from IBM Research, Stanford University and the University of Calgary in Alberta, Canada. All the major universities in the world are involved, including MIT which we mentioned, the California Institute of Technology, the University of California at Berkeley, Harvard and Cambridge. At Oxford, where Deutsch is, there is the 'Centre for Quantum Computation'. In Canada and Australia there are also Institutes for Quantum Computing.

These are just some of the English speaking centres. There are many others, like the famous Max Planck Institute in Germany, and the Universities of Vienna and Innsbruck in Austria, not to mention the important work being done in Japan and Russia. There are even journals devoted entirely to quantum computing. Nor does this include the US government agencies. One of the biggest quantum computing laboratories in the world is at Los Alamos where the first atomic bomb was built.

During the Bush administration, the US Deputy Secretary of Commerce, David Sampson, announced that the increased spending on research in the core physical sciences would be focused, "on three main areas: nanotechnology, quantum computing and the hydrogen economy."[3] That put quantum computing at the same level of importance as the hydrogen economy. The research was carried out by the main government laboratories known as the National Institute of Standards and Technology. Back in May 2005 the NATO Advanced Study Institute hosted an international conference on quantum computation. The invited lecturers came from the U.K., Germany, Canada, Russia, Poland, Austria and France.

When Do I Put Down My Deposit?

You will have gathered that quantum computing is not just some obscure 'fringe' science. So why isn't everyone putting down a deposit to get one of the first to come off the assembly line? The reason is that there is still some way to go before you can buy one over the counter. The major problem is that atoms are extremely delicate objects to try and manipulate. In order to keep the atom or particle in a state of superposition so you can use the parallel universes, you cannot bump the atom in the slightest way. The particles must remain completely isolated long enough to carry out the calculations. The most common set up is to use a thing called an iron trap which confines atomic particles within a vacuum surrounded by a magnetic field. The apparatus is then cooled to around -120° centigrade to minimise any heat which could make the superposition collapse.

Researchers are actually trying out many methods, one uses a chloroform molecule containing chlorine and hydrogen atoms which are controlled by using a variation of the medical procedure called Magnetic Resonance Imaging.

Chuang was one of the first to engineer a three qubit system, then he managed a four and five qubit system. A seven qubit system was first achieved by the Los Alamos lab. The record so far is 128 qubits achieved by Zhengbing Bian at D-Wave Systems of Vancouver, Canada, using what is called adiabatic quantum computing (MIT Technology Review, Jan. 11, 2012). It is reported (New Scientist Vol. 218 No 2918 25 May, 2013) that D-Wave Systems sold the first commercial quantum computer to the giant aircraft manufacturer Lockheed Martin for $10 million in 2011. Then in May 2013 they sold D-Wave Two to Google for an undisclosed amount. It is said to solve some problems 3 600 times faster than a conventional computer.

Where is it Taking Us?

What this seems to be telling us is that parallel universes are indisputably 'real'. It's literally quite terrifying to imagine. I mean, every atomic bit out of which we and all the objects around us are made - to the far depths of the universe - is constantly diverging into an infinite number of different universes. Who or What designed that? And where is it taking us? When we finally unlock this particular door of perception what will we find? Some strange new landscape of meaning? It's a truly wonderful mystery.

If you find it hard to actually imagine, the most helpful analogy I've found is what I call the 'Room of Mirrors'. Imagine that you are standing in a room that has all the walls covered in mirrors, as well as the ceiling and the floor – as you looked around, you would see multiple copies of yourself in all directions repeated again and again, getting fainter and fainter in the distance as though disappearing off to infinity.

That's the closest I can get to visualising the parallel universes that surround us each second of every day. Or how about each 'jiffy' of every hour? But this is not the whole picture. If you make a movement in the room of mirrors, it's copied instantly in all the images. But this is just reflecting you as a body. What's really happening is that this process of 'repetition to infinity' is happening in every one of the billions of atoms out of which you are made. Then, to cap it all, add to this Hilbert's ferocious beast – the infinite dimensional space where it's all happening! Suddenly, you find yourself wandering in some sort of incredible super sensory paradise.

One can easily just ignore the fact, much as we accept the existence of radio waves although we can't see them. But when you use your imagination to visualise what is going on, then the mental pictures that arise are simply fantastic. So foreign, so completely mysterious, so unlike the world we see with our senses. Will there be another Galileo moment when we realise it's the Earth moving? When our hunter-gatherer ancestors first wondered about the stars they could never have dreamed it would come to this. And that's only a few thousand years ago. Where will we be a million, or even a billion years from now? I would hazard a guess that we won't be locked in time or this limited three dimensional space.

References

1. Hawking, Steven. 2001. *The universe in a nutshell.* London: Bantam Press. p 80.
2. Deutsch, David. 1998. *The fabric of reality.* London: Allen Lane. p 195.
3. Sampson, David. 26th June 2006. *Let's extend the R&D tax cut.* Newsweek. p 50.

Chapter Twenty-Two
The Mystery of No Time
And No Space

God Does Not Play Dice

Just when you thought that things couldn't get worse, I'm afraid they do. If the infinite multiple parallel universes are extraordinary, then the next subject adds a whole further layer. It is one of the strangest results of the Quantum Revolution.

Although Einstein was one of the pioneers of the revolution himself, he was never happy with it. He grew up in the classical age, and both his greatest achievements, the Special and General Theories of Relativity, are classical theories. As quantum mechanics became more and more successful at explaining the atomic level, he grew more uneasy about it.

His major objection was that it wasn't exact. He didn't like the Uncertainty Principle and the way you could not measure both the speed and position of a particle at the same time. He wanted things to be certain and measurements to be definite. He is often quoted as saying that, 'God does not play dice with the universe'. So he maintained to his death that quantum theory was incomplete. There had to be a deeper, more fundamental theory that would explain all the uncertainty. He was not alone in this. Some of the other great pioneers felt the same way. Most notably Schrödinger, whose wave equation was at the heart of the Quantum Revolution.

A Famous Duel

Around 1910, Ernest Solvay, a wealthy and rather eccentric Belgian industrialist who had made his fortune manufacturing soda, decided that he wanted to try out his own unorthodox ideas in physics to the most distinguished audience he could find. This is how the famous Solvay

Conferences began. The first one was held at the Metropole Hotel in Brussels in October 1911. It was attended by Einstein, Planck, Madame Curie who won two Nobel Prizes, and many other leading figures at the time.

These conferences have been held at odd intervals ever since. It was at the fifth Solvay Conference in October 1927, that all the big names in the new Quantum Revolution were gathered. These included Einstein, Planck, Bohr, de Broglie, Heisenberg, Schrödinger and Dirac. It was at this conference that Einstein first expressed his reservations about the completeness of the new quantum theory. He didn't doubt that the theory was correct, but he thought it couldn't be the whole answer.

From here on a famous duel was fought out between Einstein and Niels Bohr, which lasted for several decades. The battles between the quantum revolutionaries and the old guard were conducted in heated debate outside the conference hall, in corridors and at breakfast and dinner as Heisenberg recalled earlier in chapter twelve. It usually started with Einstein presenting a thought experiment at breakfast which seemed to disprove quantum uncertainty. The revolutionaries led by Bohr would then gather on the barricades with counter arguments, and present them at dinner.

From 1930 onwards Einstein spent a considerable time lecturing abroad in the U.S.A. and elsewhere, and when Hitler came to power in January 1933 he decided never to return to Germany. He accepted a post at the newly established Institute for Advanced Study at Princeton, which is one of the most respected academic institutions in the world. It was here that he launched a major counter attack on the revolutionaries.

The Institute for Advanced Study

As a short break I just want to tell you a little about the Institute for Advanced Study where Einstein spent most of his working life. It was founded by a billionaire called Louis Bamberger and his sister Caroline (shades of Herschel). A sort of Utopia for very clever people. Located in 800 acres of woodland and open fields, it's a very beautiful and tranquil place. It was designed to give the students and staff the best possible environment for study.

The idea is that the professors don't actually have to teach at all, like in a normal university, so that they can get on with their research. It is entirely privately funded, though it does have close ties with the famous Princeton University nearby. It has been so successful that past students and faculty include 21 Nobel Prize winners, and 34 out of 48 Fields Medallists. The Fields Medal is the highest award in the world for mathematics because there isn't a Nobel Prize in maths.

Past members of the Institute include many of the people we have met along the way. They include Einstein of course, and his great pal Gödel, who we will be meeting shortly. Then there's Von Neumann, Robert Oppenheimer, David Bohm, Herman Weyl, Murray Gell-Mann and Freeman Dyson. It was here that Einstein continued the struggle to find an ultimate theory.

The EPR Paradox

In 1935, together with two colleagues, Boris Podolsky and Nathan Rosen, he published a paper in the May 15th issue of 'Physical Review'. It became known by the authors' initials simply as EPR. It was a landmark publication which was the subject of debate for most of the twentieth century. What the paper seemed to prove was that quantum theory contained an inevitable paradox. Schrödinger christened it 'entanglement' and the name has since stuck.

Despite the EPR paper appearing to give his own theory of the wave equation a devastating blow, he wrote excitedly to Einstein, "you have publicly caught dogmatic quantum mechanics by its throat."[1] So what was so astounding, and why was there a paradox?

What EPR shows up in quantum theory is something called 'non-locality'. By using standard quantum mechanics the equations show, as John Bell pointed out, that if you interfere with one particle of a pair, then this effect will immediately be felt by the other particle no matter how far away it is. In other words, if you were to create two identical particles, say photons of light or electrons, from the same source and let them travel further and further apart, if you then made a measurement on one of the particles, it would instantly affect the other one, no matter how far apart the two particles were. And not only this. If the particles were far enough apart so that any influence between them would have to travel faster than light, the second particle still changed, and did so instantaneously. Einstein called this 'spooky action at a distance'. He and his colleagues were convinced that this had to be impossible. After all, Einstein's relativity proved that nothing could go faster than light.

But the most shocking result was that the whole concept of 'locality' was violated. It meant that everything in the universe did not occupy a specific place (i.e. locality). This was completely absurd, and that's why it became known as the EPR Paradox. Einstein felt, as most of us would, that what happens at one place could not possibly be directly and instantly linked with what happened in another place far away from the first place. There *had* to be something wrong with quantum theory. The old guard had won a decisive victory, but in the end they were to lose the war.

Amir Aczel, a professor of mathematical science at Bentley College, Massachusetts, describes how the revolutionaries were really mad at Einstein. He says, "Neils Bohr seemed as if hit by lightning. He was in shock, confused, and he was angry."[2] At the time they couldn't refute EPR, and in the end Bohr argued that it did not stop quantum theory predicting the outcome of experiments, and therefore the EPR discovery was irrelevant. Most physicists sided with Bohr and ignored the paradox.

The Counter Attack

It took thirty years for a counter attack, by which time Einstein had died in 1955. When it came it was devastating. You will recall the maths heavyweight John Von Neumann, whose theorem seemed to demolish any chance for hidden variables and the idea of the pilot wave, and that this was taken up by John Bell, who discovered that Von Neumann's proof was not only wrong but 'simply foolish'.

Well, in 1964 Bell published a theorem known as Bell's Inequality. What he discovered was the death blow to all ideas about a local reality. 'Spooky action at a distance' was real. Niels Bohr and the revolutionaries had finally won.

The principle of locality is the way we all automatically think about the world. As I said before, things that happen in one place simply don't affect other things miles away at some other place. If I spill a cup of coffee in London it cannot possibly have any influence on someone sitting in a Starbuck's in New York.

Bell's Theorem was shattering, because it finally killed off any idea that the world could be as it appears to our senses. Henry Stapp, a renowned physicist at the prestigious Lawrence Berkeley Laboratory at the University of California, is often quoted as saying that Bell's Theorem is, 'the most profound discovery of science'. You can easily understand why. It means that everything in the universe is connected!

EPR had shown that if two particles are created together they remain 'entangled' indefinitely, so that whatever happens to one will immediately affect the other, no matter where it is in the universe. This seemed simply crazy, but Bell's Theorem proved it. So as insane as it seems, quantum theory had to be correct. What was even better, Bell's Theorem suggested that actual physical experiments might be possible to prove it beyond doubt.

John Bell came from a very modest family in Belfast, Northern Ireland. His parents couldn't afford to send him to university, so he went to the local technical college and then became a lab assistant at the famous Queens University (shades of Faraday). He so impressed the staff that money was found for him to study there. He completed his doctorate at Birmingham and then, after working at Harwell accelerator centre in England, he joined the team at CERN, contributing to its design. He was a practical physicist and ironically thought of his theoretical interest in quantum theory as just his 'hobby'!

The Aspect Experiment

It took many attempts and eighteen years more before the first defining experiment was conducted. It was carried out by Alain Aspect and colleagues in 1982 at the University of Paris in the basement of one of the laboratories. To understand what happened we need just a little more information. Bits of atoms like photons and electrons have a property called spin which is a little like the spin of a child's top, or the spin of the Earth about its axis. When they are created from the same source they always spin in opposite directions to each other according to the law which Wolfgang Pauli discovered called the Exclusion Principle.

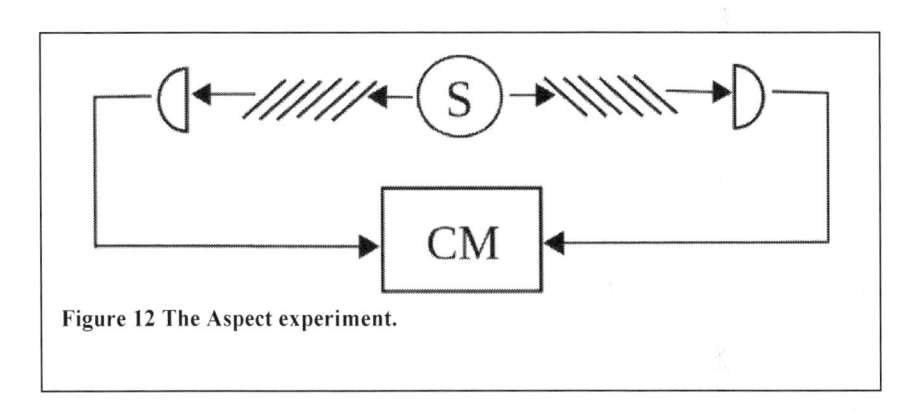

Figure 12 The Aspect experiment.

What Aspect and his team did was to send two photons with opposite spin flying apart in different directions down long vacuum tubes. Then, using very fast high frequency Switches (S = source; slanting lines are the switches and CM is the detector), they changed the spin of one of the photons. What they found was that as soon as they changed the spin of the first photon, its brother automatically obliged by altering its spin so that it remained opposite to the first photon. In fact it did this faster than the speed of light. In other words within zero time. Something, as far as we know, never before witnessed in the universe. An event which the distinguished philosopher of physics, Bernard D' Espagnat, described as 'atemporal', meaning without time.

What is happening though, does not violate Einstein's relativity as he thought it would. When the photons are first made they are 'entangled' together and part of the same superposition, so they are actually linked to each other even though they could be light years apart (because the superposition includes the whole universe). It's as though they are joined by an axle of some sort, so that when you measure the first particle to be spin 'up', the other one immediately becomes spin 'down'. You can't use this so send any messages faster than light because before you measure the first photon you don't know what its spin is going to be. That's because, as quantum theory says, it is in a great many states before you measure it.

The Andromeda Connection

Ever since Aspect's experiment many more have been conducted. The next major figure is Anton Zeilinger, professor of physics at the University of Vienna. Now, where Aspect had used the length of his laboratory, Zeilinger increased the distance between the two particles to hundreds of metres, right across part of the campus of the university.

This was then beaten by Nicholas Gisin of the University of Geneva who used fibre optic cables to send the photons a distance of 16 kilometres or 10 miles apart, and they still remained connected to each other. If a signal was passing from one particle to the other in this experiment, Aczel says, it "would have had to travel at ten million times the speed of light."[3] Phew! The longest distance so far achieved is 143 kilometres or 88 miles, between two of the Canary Islands in the Atlantic. There are plans to carry out a similar experiment using the International Space Station at a distance of 500 miles (800 kilometres). No one knows how to explain entanglement.

Not even John Bell. In an interview with Paul Davies for the BBC he said, "I don't really know....For me it's a dilemma. I think it's a deep dilemma, and the resolution of it will not be trivial; it will require a substantial change in the way we look at things."[4] D'Espagnat says, "There is no place, not even in interstellar space, where a macroscopic system can be considered to be isolated."[5] By 'macroscopic' he means things of normal size, as opposed to things at the microscopic or atomic level.

More Proof

Zeiliger has also shown that non-locality doesn't just apply to bits of atoms. He has managed to place objects, called carbon bucky balls, which are as large as seventy whole carbon atoms, in a superposition and entangle them with each other.

What this seems to mean is that everything in the universe at the atomic level is somehow connected. Either that, or another way to look at it is that space, as we currently conceive it, does not exist at this level, which is perhaps why everything is connected. If that is the case, then perhaps time does not exist either because there is no distance between things. Maybe at this level our familiar 3 space + 1 time breaks down, and gives way to the fearsome jaws of Hilbert's infinite dimensional monster? Entanglement or 'non-locality' is surely one of the most profound mysteries in the history of science. Is everything we know of just some strange instantaneous spaceless and timeless thought experiment within a grander dimensional configuration? Obviously we still have a very long road to travel. Makes you wonder though, how fantastic the realisation will be when we finally turn the key that explains non-locality.

In referring to Bell's theorem and entanglement, Aczel says, "It is amazing that such a bizarre, other-worldly property would be found mathematically, and it strengthens our belief in the transcendent power of mathematics."[6] My Oxford Concise Dictionary defines 'transcendent' as: 'existing apart from, not subject to the limitations of the material universe'. See what I mean? Bell's Theorem is yet another major triumph for maths. Even in their most creative nightmares no one would ever have come up with a universe that is timelessly connected. Is it because there is no space? This is the raw and frightening power of mathematics. And it is the subject we turn to next. But don't worry – there are still no equations!

References

1. Aczel, Amir. 2002. *Entanglement the greatest mystery in physics.* Chichester: John Wiley & Son. p 119.
2. ibid., p 118.
3. ibid., p 237.
4. Davies, P. and Brown, J. 1986. *The ghost in the atom.* Cambridge: Cambridge University Press. p 48.
5. D' Espagnat, Bernard. 1989. *Reality and the physicist.* Translated by Whitehouse, J., Cambridge: Cambridge University Press. p136.
6. Op. cit., Aczel, A. p 252.

Chapter Twenty-Three
The Mystery Beyond the Multiverse

Early Champions

The greatest mystery in the universe at this 21st century moment in time, the mystery that has made our information age possible, is the mystery of mathematics. Without it science, as we know it, wouldn't exist and we'd still think the Sun went round the Earth. One of my favourite quotes about the mystery of mathematics comes from Schrödinger who once said:

"Mathematical truth is timeless, it does not come into being when we discover it."[1]

Timeless?...Something that exists before we discover it? Makes you wonder if it's something not of this world. Something entirely alien to the normal stuff the universe is made of.

All along this road that we have been travelling, the power of mathematics to describe the real world has been a strong undercurrent to everything we have looked at. It has been a major theme that I've mentioned at every turn in the story. It is now time to look at it a bit closer. Time for us to confront the mystery beyond the multiverse!

Most people think that maths is something created inside our heads. But, as we've discovered along the way, some of the most famous mathematicians and scientists throughout history have believed that maths is independent of the human mind, and although you can't see it or touch it, or smell it for that matter, it is actually the blue-print for the 'everything' we call reality. It seems to be what the physical world, which we see with our eyes, is actually made out of. What I want to do is to review all the evidence.

THE FINAL MYSTERY

Perhaps the first to have these ideas was Pythagoras and his followers. They believed that everything in the universe was made from numbers and that reality at its deepest level was mathematical. They discovered, by analysing the vibration of strings on musical instruments, the mathematical sequences that lay behind music and harmony. This led them to believe that the universe could be explained by numbers. There is a wonderful irony here because, as we saw earlier, the leading candidate for a Theory of Everything to day is called String Theory which pictures bits of atoms arising from the vibrations of tiny strings at the Planck scale.

Plato is the next major figure. In a similar way to the Pythagoreans he believed that maths is a non-physical ideal world that lies behind our ordinary every day reality, and that our experience of everything is just a poor and imperfect copy of this ideal world. People who believe this are still called Platonists.

Davis and Hersh, who wrote a famous classic about maths, confirm this dark secret that lies at the heart of science. They say, "Platonism was and is believed by (nearly) all, mathematicians. But, like an underground religion, it is observed in private and rarely mentioned in public."[2] I should say that Sir Roger Penrose, whom we've met before, is one of the worlds' leading mathematicians. He is the Emeritus Rouseball Professor of Mathematics at Oxford. Back in 1990 he was courageous enough to break ranks with his colleagues and 'come out' so to speak, in his first popular science book, and admit in public that he is a Platonist. Had it been another era he would almost certainly have been burned at the stake. In a recent article for New Scientist he says, "the truths that mathematicians seek are, in a clear sense, already 'there', and mathematical research can be compared with archaeology; the mathematicians job is to seek out these truths as a task of discovery rather than one of invention."[3]

In a famous passage in Plato's book 'The Republic' he describes how we are all like slaves chained to a wall inside a cave. There is a fire burning in the cave and it throws shadows on the wall opposite. What we see is just the shadows, and because that is all we can see we believe in our ignorance that the shadows are the real world.

Galileo Again

Although there were many in between, it is Galileo who stands out next because he is acknowledged as the father of modern science. He is famous for saying that the language of nature is written in mathematics, and that God is a mathematician. He discovered that the law governing how things fall was completely different from what Aristotle had maintained, and Aristotle had been the accepted dogma for some eighteen centuries.

Aristotle said that objects fall at a speed that is proportional to their weight. In other words, that heavier things fall faster than light things which is exactly what your senses tell you as I said before, if you hold a golf ball in one hand and a solid iron shot putt in the other. But with a clever thought experiment Galileo asked those who supported Aristotle to imagine a brick falling from a high tower. If the brick was to crack in half on its way down it wouldn't suddenly slow down to half the speed because the two halves were now lighter than the whole brick had been. It seems amazing that no one had thought of this before.

By doing actual experiments, and using his own pulse to count time, he came across the 'time squared' law. He found that things didn't fall at a constant speed as Aristotle had said they did, they accelerated. If something was falling at a constant speed it would travel the same distance in each second. So, say it fell 5 metres at the end of the first second, and another 5 metres after the 2nd second; that would make it fall 10 metres in all after 2 seconds. But Galileo found that if an object fell 5 metres at the end of the 1st second, then at the end of the 2nd second it would have travelled by the time squared. In other words $2^2 = 4$ x 5 metres = 20 metres! Double what Aristotle thought. He had discovered a law of nature that no one knew existed. It wasn't a law carved in stone. It was a purely mathematical law. A law that explained how things behave.

Miracles

Next, I'm going to jump to modern times to an iconic paper published by none other than Eugene Wigner in 1960, which he called in typical physics-speak, 'The Unreasonable Effectiveness of Mathematics in the Natural Sciences'. To the average person this doesn't sound too remarkable, but in an essay of about fifteen pages he uses the word 'miracle' no less than eight times – and this is a physicist who won the Nobel Prize! The first point he says, "is that the enormous usefulness of mathematics in the natural sciences is something bordering on the mysterious and that there is no rational explanation for it."[4] Mysterious...no rational explanation? And later, "It is difficult to avoid the impression that a miracle confronts us here."[5] This is the miracle of maths we are exploring – the greatest mystery in the universe.

So what examples does he give about the mysterious power of maths to describe the world? He discusses Galileo's law of free falling bodies, then moves on to Newton's discovery of gravity, which he describes as a monumental example because it has been proved to be accurate to less than ten thousandth of a percent. It is so accurate that it is still the main tool used to plan all space missions. He then moves on to quantum theory and how it is able to describe a helium atom with an accuracy of one part in ten million!

An important bit, from our point of view, comes towards the end where he discusses semiconductors and insulators, and how the electrical resistance of insulators maybe 10^{26} times greater than metals. This is a huge number, so big and therefore so accurate, that he is moved to say, "there is no experimental evidence to show that the resistance is not infinite."[6] He ends the essay by saying the miracle of maths, "is a wonderful gift which we neither understand nor deserve."[7]

The Mystery in Action

It still amazes me to think that Newton managed to discover how gravity works and apply this to how the planets orbit the sun, and things so utterly different as falling apples and the moon creating the ocean tides

...using nothing more sophisticated than a feather quill and ink made from soot and water! It's quite preposterous when you think about it. So powerful is this mathematics that it can explain things like the rotation of galaxies and black holes. Not to mention the actual behavior of the universe including objects a thousand million light years away! No wonder Wigner thought it a miracle.

Another example I like took place in 1846 when a Cornishman and a Frenchman, who had never heard of each other, and using different methods, both discovered the planet Neptune, just by manipulating squiggles on a piece of paper, and probably by candlelight! At the stroke of a pen (literally), they discovered that our Solar system is far bigger than we thought it was. Only later did telescopes find it. From studying the movement of the then known planets, they figured there must be another one, and by using mathematics they were able to pinpoint where it should be. When they turned their telescopes on the spot they found it, exactly where it had been predicted!

Another landmark in the power of maths came as we saw earlier, when Maxwell discovered the equations that showed electricity and magnetism to be the same thing. I mean lightning and fridge magnets? You could hardly think of two more different things. And then, if you remember, how he held his breath when he first realized that they were the same thing as light! Nor does it stop there, the equations also revealed the existence of radio waves and the rest of the electromagnetic spectrum which were only discovered later. From...squiggles on paper?

We have already covered Einstein but, since we are concentrating on the mystery of mathematics, I want to remind us again of two aspects of the mystery. Firstly, there are many occasions where mathematicians have discovered new kinds of maths that then lie unnoticed for sometimes a hundred years or more until miraculously much later, physicists in exploring how the world works, find that it behaves according to this forgotten maths. The second is our main theme, which is the miracle of how mathematics allows us to discover things about the world that we would never have guessed otherwise, like the fact that electricity and magnetism are the same thing as light.

THE FINAL MYSTERY

Einstein provides examples of both. He didn't know the mathematics of curved surfaces, so he had to get his old pal Marcel Grossman from his days at the patent office to teach it to him. It was a new kind of geometry discovered by Riemann a hundred years earlier as a purely academic exercise, as we saw in chapter eight. But it turned out to prove not only that space and time are the same thing – a big enough miracle in itself – but also that it could describe not just the solar system, it could describe the whole universe! If you think about it too long it can get a bit scary. How can patterns that just follow an internal logic using arbitrary symbols possibly describe something as gigantic as the observable universe?

Next, the equations kept telling him that the universe had to be either expanding or contracting. It could not be static and stationary. Yet everyone who had ever looked at the night sky and thought about the universe, had never dreamed that it could possibly be *moving*. Einstein was no different. He thought the equations *had* to be wrong. So he cheated. He told everyone he was doing it of course, and he added a bit to the equations to keep the universe still, and called it the Cosmological Constant.

Just over ten years later Hubble was able to measure that the universe was indeed expanding! This is surely one of the greatest discoveries in all of science. If Einstein had only believed in the equations, how much bigger would his legacy have been? He later admitted that it was the greatest blunder of his life. I want to add an anecdote here, just briefly, because it neatly illustrates what we are talking about. In 1931 Einstein visited the U.S.A. When he was in California, Hubble invited him and his wife Elsa to have a look at the telescope on Mount Wilson where he had discovered that the universe was expanding. When it was explained to Elsa that the equipment was used to determine the size and shape of the universe, she is said to have replied, "that's nothing. My husband does it on the back of an old envelope!"

The third thing about the mystery of mathematics is that it can arrive at the same conclusions in describing the world by quite different routes.

This happened as we saw with the discovery of Neptune. What it seems to suggest is that maths might be much bigger than our minds which are obviously tied to what is in the physical universe. We will get to this later, but Heisenberg and Schrödinger give us another excellent example of this bit of the mystery.

In 1927 when Heisenberg escaped to the island of Heligoland to cure his hay fever, he came up with matrix mechanics to describe how the atom behaved, only to discover when he got home that mathematicians had been playing with matrixes for years. Weinberg says, "this is one example of the spooky ability of mathematicians to anticipate structures that are relative to the real world."[8] It's an example of the same thing as Einstein with Riemann geometry. But the third bit of the mystery comes in when Schrödinger goes skiing in the mountains with his girl friend and discovers his famous wave equation, which describes the same thing as Heisenberg's matrix mechanics, but by a quite different kind of mathematics.

Do you remember Q.E.D., the theory that got Feynman his Nobel Prize? Q.E.D, the most accurate theory so far in the history of science? The same theory was arrived at by two other people using a different kind of mathematics. The one was Julian Schwinger if you remember, and the other was Sin-Itiro Tomanaga, who was working quite independently in the oblivion of war-torn Japan. They shared the same Nobel Prize with Feynman in 1965, but only after Freeman Dyson at the Institute for Advanced Study was able to demonstrate that they were one and the same theory. Unfortunately he didn't get a Nobel Prize.

Because we are looking at the mathematics of the Quantum Revolution, which is as fundamental as it gets, then the realization that there can be three different ways to describe it, begins to suggest that mathematics might be bigger, not just than our minds, but also perhaps of physical reality itself. David Deutsch makes the point that the maths we know is limited by our physical brains, and that this is just a tiny part of the whole landscape of mathematics. "The comprehensible mathematical truths are precisely the infinitesimal minority which happen to correspond exactly to some physical truth."[9] In other words maths is gigantic in capacity compared to the physical universe. In fact it is not restricted by anything like size because it exists before anything has size – before space exists. That is quite frightening.

Let's return briefly to what is perhaps one of the most spectacular examples of the power of mathematics to describe our world. It comes from Paul Dirac who Einstein once described as dizzyingly brilliant, because he didn't think he was very good at maths himself. The same person, if you remember, who couldn't understand how Heisenberg knew that a girl was nice before he danced with her. When he was trying to combine Special Relativity with Quantum Theory, the equations kept telling him that there was a particle never seen before which was exactly the same as an electron but with a positive charge. Try as he might he couldn't get rid of it. He became really frustrated with it, and in the end accepted it just to make the equations work.

He had discovered antimatter. Something that no human mind had ever minutely conceived of, not even in dreams. Yet a few years later as you'll recall, Carl Anderson, doing experiments with high altitude balloons, detected the same particles that became known as positrons. Since then it has been proved resoundingly through rigorous experiments that every known particle has an antimatter partner, and antimatter is now regularly produced in accelerator experiments. Yet this is, as it happens, just a barely important side-show. It is now accepted that, microseconds after the Big Bang, a contest between matter and antimatter decided the actual structure of our universe and hence the emergence of our consciousness. Steven Weinberg says that the, "excess of matter over antimatter is one of the key initial conditions that determined the future development of the universe".[10] Discovered, yes… by just using squiggles on a piece of paper!

This is the great mystery of mathematics written large. The story goes on though. Let's think again about de Broglie, hammered into silence by the bully Von Neumann. He demonstrated mathematically that apparently solid bits of atoms like electrons and protons were also waves! People thought, 'how can bits also be waves?' Yet early experiments proved that the maths was right. Conservative scientists would never have dreamed it. It was a total revolution.

How about Hideki Yukawa's discovery of the pion that carries the strong nuclear force, or Murray Gell-Mann's prediction of the omega-minus? Then take the story of Weinberg and Salam uniting two fundamental forces of nature (imagine having that on your CV by the way?), and then the W^+, W^- and Z^0 being discovered by experiment. What about the quark controversy where no one *wanted* there to be yet another level deeper than the proton and neutron? At first they were convinced that quarks were just a convenient mathematical device which could explain the results of experiments, but they turned out to be real. When someone suggested at a conference that quarks might actually be real, it was reported that Gell-Mann got up and walked out. Like Weinberg said, it's not that they take their theories *too* seriously, the evidence is that they don't take them seriously enough. And then there is the staggering triumph of Bell's Theorem which suggests that everything in the universe is connected!

The Great Internet Prime Search

It's time to look a bit closer at simple numbers. The best way to actually see the mystery of mathematics at work is by looking at prime numbers. We all come across them in school, but they never explain the 'pure magic' involved. Here it is. Prime numbers are numbers that can only be divided by 1 and themselves. They are all odd except for 2 which can only be divided by itself and 1. Here are the first dozen: 2, 3, 5, 7, 11, 13, 17, 19, 23, 29, 31, 37. Take the number 3 for example. You can divide it by either 1 or 3; nothing else goes into 3 without there being a remainder left over. This is also the case for the next prime number which is 5. But if you go on to 6, then you can have 1x6 = 6, but you can also have 2x3 = 6, so 6 isn't a prime. If you consider 10 then it can be 1x10 = 10, or 2x5 = 10. But 11 can only be 1x11 = 11. Nothing else will divide exactly.

Or take 28. It can be:
1x 28 = 28
4x7 = 28

14x2 = 28. So 28 isn't a prime either. But the next number, 29, can only be 1x29 = 29. It seems strangely out of our control.

They start out being quite frequent, but as the numbers get bigger the number of primes reduces, but not in a way that anyone can predict. Literally, when it gets to bigger numbers there is no way to predict where the next one will pop up. They occur randomly. That's why the military and government use them in secret codes. Now, if maths *was* created by the human mind we would surely be able to predict the next prime number, after all it's *us* that is creating the numbers, right?

The truth is that we haven't a blind clue about the next prime number. They have tried them out using supercomputers. The biggest prime discovered to date is over 17 million digits long, and there is still no pattern to predict where the next one will happen. Surely, if the human mind *is* creating maths, it would know when the next prime number would come up, but we have no idea. So maths certainly seems to be independent of the human mind.

There are now thousands of enthusiasts all over the world using their PC's to take part in an internet project to find the next prime number. Interestingly, it is the only game in town that can never end. Why? Because way back in 300 BC, Euclid proved that there must be an infinite number of primes!

The Voice of the Revolution

About the amazing power of mathematics, David Layzer, Professor of Astrophysics at Harvard University, wonders, "why it is that the regularities which lie deep beneath the outward appearance of our physical world are actually mathematical, and even more mysterious; why they should be in the least bit accessible to the human mind." When you think about it: why *should* the human mind have access to this silent realm? Is it telling us something profound about our connection to reality? This seems to be a deep secret in itself. He goes on to say, "these are the great mysteries at the heart of humankind's most sustained and successful rational enterprise."[11]

John Barrow, whom we met in the introduction, is another big international name who is the Professor of Mathematical Sciences at Cambridge University. He confirms the dark secret we are uncovering that lies at the heart of science. This is what he says about mathematics, it's quite electrifying: "A mystery lurks beneath the magic carpet of science, something that scientists have not been telling, something too shocking to mention."[12] At the root of the success of science he says, "there lies a deeply 'religious' belief – a belief in an unseen and perfect transcendental world that controls us in an unexplained way, yet upon which we seem to exert no influence whatsoever."[13] See what I mean? Makes the hair on the back of your neck stand on end!

Don't waste your time wondering about the paranormal, or telepathy, or UFO's. The biggest mystery by far in this universe, and the multiverse beyond, is quite simply – what is the nature of mathematics?

So you might well then ask, where *is* maths? Where is it actually *located* ? One of the most chilling aspects of its mystery is that it exists outside both space and time. This can surely only mean that it is in some way bigger than anything 'physical', like a multiverse. Bigger even than an *infinite* number of physical universes! That is why discovering the nature of mathematics has got to be, by definition, the greatest revolution ever to confront humanity. Of all the revolutions in human consciousness we have encountered on our journey, this has to be the greatest. We have discovered that there is something, a phenomenon if you like – that actually 'exists', but one which is not physical and is beyond both time and space. And we are not talking about ethereal spirits or angels here, we are talking about where our physical reality comes from at the deepest level.

Let's explore it a bit further. You'll recall that when Einstein was asked to be the first President of the new state of Israel after the Second World War he turned it down, saying that politics was for the present, whereas an equation was for eternity….and he meant – eternity. What he was saying is that, once an equation is discovered to be correct, it will always be correct, not just now but forever, and not just here on Earth. It will be just as true on Alpha Centauri or the other side of the universe. It is not limited by space. Nor is it limited by time. It still gives me a shiver up the back of my spine.

Once discovered it will always be correct, and its truth, its logical correctness, existed before it was discovered. As Schrödinger said, mathematical truth is timeless. It does not come into being when we discover it.

Scientists believe that the best way to communicate with intelligent alien life when we find it (or it finds us), will be through mathematics. The squiggles they use might be different from the ones we use, but the underlying truth is the same throughout the universe. Just as we know what the number 9 means, a Roman would also know, except he would write it as 'ix'.

Max Tegmark, a Professor of Physics at the famous Massachusetts Institute of Technology puts it like this. "There is nothing fuzzy about mathematical structures. They are 'out there' in the sense that mathematicians discover them rather than create them, and that contemplative alien civilisations would find the same structures, (a theorem is true regardless of whether it is proven by a human, a computer or an alien)." He continues, "a mathematical structure cannot change – it is an abstract, immutable entity existing outside of space and time."[14]

Steven Weinberg has said: "Matter thus loses its central role in physics: all that is left is principles of symmetry…" In a footnote he goes on to say about the properties of quantum particles, "from this point of view, momentum and spin are what they are because of the symmetry of the laws of nature."[15] According to a classification by Eugene Wigner the symmetries of the laws of physics actually determine the properties of the particles found in nature, and thus the whole universe including us. But a symmetry principle is a purely abstract entity, a mathematical object of pure information. It has no content other than its ability to prescribe how reality is constructed.

John Barrow is even more explicit. He says, "The reason why mathematics is so successful in describing the way the world works is because the world *is* at root mathematical"[16] (his emphasis). And again in another source he says, "If one takes matter to pieces and probes to the root of what those pieces 'are' then ultimately we can say nothing more than that they are mathematics."[17] Is this a rumble of thunder foretelling the arrival of the revolution?

Peter Atkins, an Oxford professor, speculates that perhaps, "the deep structure of the world is mathematics: the universe, all it contains, *is* mathematics, nothing but mathematics, and physical reality is an awesome manifestation of mathematics."[18] Getting even deeper perhaps, he says elsewhere, "It is possibly not too extravagant to claim that the answer to the question of why mathematics works will be the final answer to all questions of being"[19]. The final answer to all questions of being? Another hair-raiser!

Recently Nima Arkani-Hamed at the Institute for Advanced Study who has previously held professorships at both Berkeley and Harvard, has discovered geometrical structures called Amplituhedrons. Based on Twistor theory developed by Roger Penrose and Andrew Hodges at Oxford university, these new structures not only simplify the standard approach discovered by Richard Feynman, but at a deeper level they appear to confirm that space and time are not fundamental but are derived from geometry.

David Deutsch, who is the founder of Quantum Information Science and a champion of parallel universes, is another Ché Guevara of science. He is very straight forward, even quite blunt about the nature of mathematics. This is what he says: "Mathematical entities are part of the fabric of reality because they are complex and autonomous…although they are by definition intangible, they exist objectively and have properties that are independent of the laws of physics."[20] Independent of the laws of physics? That's like saying that mathematics is independent of everything 'physical'. It's one of my favourite quotes, it's so explicit. I should say at this point that, with the exception of Roger Penrose, most of these scientists would not consider themselves Platonists. In the philosophy of mathematics they are probably best thought of as Structural Realists. They don't believe in some 'ideal realm' separate from physical reality. They believe physical reality is at base mathematical.

Looking Through the Wrong End of the Telescope

When you look at the pattern of seeds in the head of a sunflower, or the shape of ferns, or the petals of a flower, you can see an obvious symmetry which is 'out there'. Maths can reproduce this symmetry with numbers. It's called a Fibonacci Sequence after the Renaissance Italian mathematician who introduced Western civilisation to the Indo-Arabic number system. But that's just a code to describe what is already existing in the world of Nature. The elaborate symmetry of a simple pine cone and the fact that electricity and magnetism are the same thing as light, existed long, long, long before evolution came up with the human mind. Those who think we create mathematics in our heads are clearly looking through the wrong end of the telescope. As the internationally renowned mathematician Amir Aczel says, "numbers, therefore, cannot be our invention. They are entities about which we learn fascinating new things all the time." He goes on to say, "Numbers do exist, and their existence, I believe, is independent of people." He continues, "In another universe, one without people and without anything we recognise from our own universe, numbers will still exist. And these numbers are infinite."[21] The fact that numbers are infinite is crucial to where we are going.

The discovery of the Mandelbrot Set, those never-ending self-repeating patterns produced by an equation on a computer that were once popular as wall posters, seem to demonstrate how Nature appears to create the shapes of leaves and trees. More exciting still, though, is that when you look at the Mandelbrot fractal you can take a tiny section and then magnify it up say ten times, and within the section you chose to magnify there appears the same pattern on a smaller scale. If you then take a bit of that pattern and magnify it again, the same bit of pattern repeats itself, again and again. And the deeper you go, it *still* repeats itself as you descend towards infinity.

THE MYSTERY BEYOND THE MULTIVERSE

This is what Roger Penrose says about the Madelbrot Set. "The Mandelbrot set was certainly no invention of any human mind. The set is just objectively there in the mathematics itself. If it has meaning to assign an actual existence to the Mandelbrot set, then that existence is not within our minds, for no one can fully comprehend the set's endless variety and unlimited complication...Its existence can only be within the Platonic world of mathematical forms." To explain further, he continues like this. "The mathematical forms of Plato's world clearly do not have the same kind of existence as do ordinary physical objects such as tables and chairs. They do not have spatial locations; nor do they exist in time. Objective mathematical notions must be thought of as timeless entities and are not to be regarded as being conjured into existence at the moment that they are first humanly perceived." Later he says: "Those designs were already 'in existence' since the beginning of time, in the potential timeless sense that they would necessarily be revealed precisely in the form that we perceive them today, no matter at what time or in what location some perceiving being might have chosen to examine them."[22] Pretty breathtaking stuff.

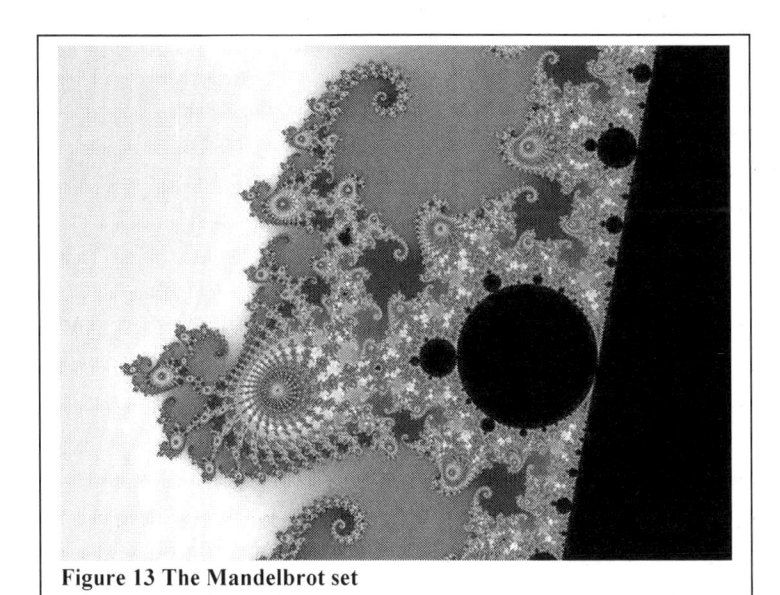

Figure 13 The Mandelbrot set

Here is some more. In a memorably poetic turn of phrase for a mathematical scientist, John Barrow says, "The beautiful subject of fractals has revealed itself to underwrite the whole spectrum of natural phenomena – from the clustering of galaxies to the crystalline structure of snowflakes."[23] What could be more wonderful than this realisation? It dwarfs into insignificance every human drama ever written and reduces all our religious and ideological conflicts to trivia. It explains with a single brush stroke the whole nature of our reality. Interestingly, recent evidence from a satellite called the Sloan Digital Sky Survey suggests that our universe might itself be a fractal.

The Revolution

When you next look at a head of broccoli, stop and imagine for a moment the equation that is staring you back in the face. Will the final Theory of Everything be like this? Will there be an equation that can grow an entire universe like ours, with conscious beings who can think about it? Hawking once said that discovering this set of equations would be the ultimate triumph of human reason. But, with respect, it seems to me it would just be another beginning.

As we saw earlier, any proper Theory of Everything, would have to be able to describe not only our universe, but the infinite other universes that make up the multiverse. Nor will this be sufficient either, because it will also have to account for the mystery of mathematics itself. Nor am I alone in saying this. Many scientists are now coming to this realisation. Paul Benioff of the Argonne National Laboratory and a pioneer in quantum computing argues that, "a final theory of everything should not merely unify all of physics, but should also provide a common explanation for physics and mathematics."[24]

John Barrow says it much more boldly. "The meaning of mathematics will emerge as a key question that must eventually be answered in any quest for a fundamental understanding of the physical world."[25]

Stephen Hawking, perhaps the most famous scientist of his time, said about the mystery of mathematics: "Even if there is only one possible unified theory, it is just a set of rules and equations. What is it that breathes fire into the equations and makes a universe for them to describe?"[26]

Describing mathematics as a code, Marcus du Sautoy, the Simonyi Professor for the Public Understanding of Science, and Professor of Mathematics at the University of Oxford, says, "the fact that the code provides such a successful description of nature is for many one of the greatest mysteries of science."[27] He goes on to say later, "When we start to look closely at all this complexity, surprising patterns begin to emerge. It is these patterns that I believe point to an underlying code at the very heart of existence, and that controls not only our world and everything in it, but even us."[28]

Alain Connes, a very distinguished French mathematician (Fields Medal, 1982), says: "I hold on the one hand that there exists, independently of the human mind, a raw and immutable mathematical reality: and, on the other hand, that as human beings we have access to it only by means of our brain...I therefore dissociate mathematical reality from the tool we have for exploring it."[29] Later he asserts further: "I believe that not only does this physical reality exist, but that there also exists a mathematical reality."[30]

So it seems that everything comes from mathematics. Something we don't understand. Something that exists outside of space and time, and something more comprehensive than the universes it creates. It is the only thing we know of that is not derived from something else. This must surely be the deepest mystery ever to confront our smudge of consciousness?

This first quarter of the 21st century is surely the 'now' time. The time for science to come clean. Time for it to reveal this secret - 'too shocking to mention'. Time for this 'underground religion' to see the light of day. Time to man the barricades for a revolution far greater even than the Quantum Revolution.

References

1. Schrödinger, Erwin. *What is life and mind and matter?* Cambridge: Cambridge University Press. p 154.
2. Davis, P., and Hersh, R. 1981.*The mathematical experience.* London: Penguin. p 339.
3. Penrose, Roger. 18th Nov. 2006. *What is reality.* New Scientist No 2578. p 38.
4. Wigner, Eugene. 1991. *The unreasonable effectiveness of mathematics in the natural sciences.* In *The world treasury of physics, astronomy and mathematics.* Edited by Ferris, Timothy. London: Little Brown & Co. p 527.
5. Ibid., p 533.
6. Ibid., p 540.
7. Ibid., p 540.8
8. Weinberg, Steven. 1993. *Dreams of a final theory.* London: Vintage. p 52.
9. Deutsch, David. 1998. *The fabric of reality.* London: Peguin. p 255.
10. Weinberg, Steven. 2003. *Facing up.* London: Harvard University Press. p 73.
11. Layzer, David. 1990. *Cosmogenesis.* Oxford: Oxford University Press. p 14.
12. Barrow, John. 1992. *Pi in the sky.* London: Penguin. p 1.
13. Ibid., p 1.
14. Tegmark, Max. 2004. *Parallel universes.* In *Science and ultimate reality.* Edited by: Barrow, J., Davies, P. and Charles Harper, Jr. Cambridge: Cambridge University Press. pp 480 – 482.
15. Weinberg, S., op. cit. p 111.
16. Barrow, John. 1991. *Theories of everything.* London: Vintage. p 183.
17. Barrow, John. 1999. *Between inner space and outer space.* Oxford: Oxford university Press. p 9.
18. Atkins, Peter. 2003. *Galileo's finger.* Oxford: Oxford University Press. p 355.

19. Atkins, Peter. 1994. *Creation revisited.* Oxford: Oxford University Press. p 101.
20. Deutsch, David., op.cit. 1998. p 255.
21. Aczel, Amir. 2000. *The mystery of the aleph.* London: W.S.P. p 226.
22. Penrose, Roger. 2005. *The road to reality.* London: Vintage Books. p 16-17.
23. Barrow, J., op. cit. 1999. p 87.
24. Davies, Paul. 2006. *The goldilocks enigma.* London: Allen Lane. p 272.
25. Barrow, J., op. cit. 1992. p viii.
26. Hawking, Stephen. 1988. *A brief history of time.* New York. Bantam Press. p 174.
27. Du Sautoy, M. (Presenter), & Cooter, S. (Director). (Oct. 2011). *The code,* [television series episode 1]. In P. Leonard (Producer). London, BBC Studios.
28. Ibid., episode 3.
29. Changeux, Jean-Pierre. and Connes, Alain. (Translated by DeBevoise, M. B.) 1995. Conversations on mind, matter, and mathematics. Princeton: Princeton Uiversity Press p 26.
30. Ibid., p 200.

Chapter Twenty-Four
Are There Limits To Reason?

Slippery Things

Now we come to the really exciting bit! I mean, the *most* exciting bit. We have been talking in a guarded sort of way about those things we call infinities. We've discovered that they are everywhere, from the infinite parallel universes at the atomic level to the infinite mega universes of the multiverse. In fact, we seem to be surrounded by infinities in every direction as Freeman Dyson once suggested. John Barrow makes the point that: "Mathematicians have also had to face up to the reality of infinity. The issue was a big one, one of the biggest that mathematicians have ever faced."[1]

I once said that they are the most daunting objects known to us. I meant it. There can't be anything bigger or more extensive than something that goes on and on forever. By its very meaning an infinity is the biggest object that the human mind can comprehend. In the 11th century AD, St. Anselm of Canterbury came up with a definition of God, 'as that than which no greater can be conceived'. That's what an infinity is.

The infinity that most of us are used to is that of the natural numbers: 1,2,3,4,5,6,7,8,9,10,11… a hundred billion… and so on. There is no 'biggest' number. That's because you can always just add another number to the one you just thought was the biggest. So the first one wasn't the biggest! In a similar way you can ask, what is the smallest possible number that is not actually zero? Let's call it 'itsibitsi' number. Now divide it by 2. You have just produced a new number that is still bigger than zero but smaller than itsibitsi. So itsibitsi isn't the smallest number! Strange isn't it? Eli Maor, a well known Israeli mathematician, says, "To the mathematician, infinity *is* a reality [his emphasis]. In fact, mathematics could hardly exist without it, for it is inherent already in the counting numbers, which form the basis of practically all of mathematics"[2]

This is how deeply entwined infinity is with maths. Yet we have already seen all the evidence for mathematics being the blueprint of our reality. So infinities appear to be basic to what makes our world. It's a bit unnerving when you think about it, but it also presents us with a tremendous adventure.

Then there is the remarkable story of 'pi', which is just the simple ratio of a circle's circumference to its diameter, please don't switch off now, everything that follows is very obvious. In other words, how much longer is the line around the circle, than the line that divides it in half. It's an actual physical measurement, but one which can never ever be an exact amount no matter how big the circle. It always comes out as an infinite decimal expansion. Pi = 3.14159265…and so on, forever. You never come to a definite 'whole number' answer. Why is that? After all, this world is meant to be a physical place. Surely it has to have an exact answer?

On the other hand, perhaps this is a kind of hidden proof that infinities are part of the 'real' world. Supercomputer calculations have determined pi to over one *trillion* digits, and it hasn't ended. And like the prime numbers, no pattern in the digits has ever been found. Yet pi has proved to be enormously useful in physics, and it appears on a routine basis in equations describing the most fundamental laws of the universe.

Similarly, all engineers rely on approximations to infinity in their calculations every day of their working lives because it gives them the greatest possible accuracy in designing things, like aircraft and bridges and shopping malls, and they are quite happy with it because it works!

Another piece of magic which contains an infinite expansion is the $\sqrt{2}$ which is equal to: 1.414213562…and so on forever. In physical terms it could hardly be more simple. It is just the length of the diagonal of an ordinary square with sides that are equal to 1. But no matter how big or tiny or gigantic you draw the square, the diagonal will never be equal to a 'whole' number! It seems very mysterious that something so ordinary and simple can hold within it an infinity that seems to be bigger than the universe. Why *is* that? I should say that the square root of all prime numbers is also like this.

Infinities are quite slippery things. Take a straight line. It's about the simplest and least complicated thing we can possibly think of. I mean, just an ordinary plain line. Yet it is actually shrouded in fabulous mystery. I feel sure you will not want to believe it, but consider this: we know that any and every line is made up of an infinite number of 'points' – that's what 'makes' the line. This means that a single line a million miles (1.6 million kilometres) long is made up of an infinite number of points, but so is a line just one inch (2.5 centimetres) long. See what I mean? I do hope you are enjoying this – it gets even better!

Infinities seem to define everything that exists in a way our minds can't yet comprehend. They appear to point to something that is not limited by 'space', like the length of a line. And also something that is not limited by 'time'. Paul Davies (Director of the Centre for Fundamental Concepts in Science, Arizona State University), reminds us that, "there are no more moments in all of eternity than there are in, say, one minute. In both cases there is an infinite number."[3]

Those who think that these infinities are just something created by us should consider 'now'. As I said earlier, is the 'present moment' perhaps one second, or is it half a second? Perhaps it's a tenth of a second, or maybe a hundredth? Or is it a jiffy, the Planck time of: 1/000 000 000 000 000 000 000 000 000 000 000 000 000th of a second? A jiffy is just as real as a second. It seems to be telling us something very fundamental about all of reality. All these infinities, all the ones we've been looking at, seem to suggest that every possible option actually exists. It's as though our reality, our physical world if you like, is just a sort of shimmering arrangement of logical mathematical consistency, "a quivering slice of mathematical stability [lost] in an infinite ocean of all possible possibility."[4]

It's as though finding a Theory of Everything may be a matter of discovering why our particular mathematical blueprint predominated over all the other possible alternatives, despite the fact that all the others do actually exist all around us. But like radio waves, we can't see them. That's a bit frightening, but it also pretty exciting.

We also know that our particular shimmering arrangement is certainly not perfect. Perhaps in the far future we may be able to manipulate the mathematical options to create a less flawed universe. This sounds naïve now, but recall that, were it not for the evolution of consciousness from stardust, then thinking this would not be possible. Nor could the stardust have dreamed it. Could this be the ultimate purpose of consciousness?

Live Ones

But, as we have seen, infinities don't just happen in the world of numbers. Not only are we surrounded by infinities at the universal and microscopic levels, but real live ones, so to speak, exist in the heart of black holes. Places where matter becomes infinitely dense and in so doing actually squeezes space and time out of existence. It is now thought that a black hole resides at the centre of each galaxy including the Milky Way.

There are also less hazardous infinities than a black hole which swallows everything in sight. Ones that are also very real, like the one caused by the cosmic speed limit. Remember how every day in the big particle accelerators they move tiny bits of atoms to 99.999% of the speed of light. But to reach 100%, or the *actual* speed of light, would require more energy than there is in the entire universe. That's because as anything approaches the speed of light it needs more and more energy to accelerate it. To reach the speed of light it would take an 'infinite' amount of energy.

Then let's take something as ordinary as the coastline of Britain. You can measure it on a map of course, but if you want to measure it exactly you have to go down to the beach with a digital theodolite. And even that is not enough, you will need a magnifying glass to measure each grain of sand, and then an electron microscope to measure each atom, and so on.

Computers can also handle 'real' infinities. Marcus Chown, a well known writer and broadcaster, says, "real computers don't just perform finite computations, doing one or a few things, and then halt. They can also carry out infinite computations, producing an infinite series of results."[5] Veronica Becher of Buenos Aires University explains that, "many computer applications are designed to produce an infinite amount of output."[6] The examples given are web browsers such as Netscape and operating systems like Windows. They use something called Binary Floating-Point Arithmetic that uses infinities to avoid errors. So although most of us don't realise it, infinities are part of our everyday lives.

Georg Cantor

It was a famous (Russian born) German mathematician, Georg Cantor, in the middle of the 19th century who was the first to discover that there are different kinds of infinity. Not just a few either. He actually proved that there are an infinite number of infinities! According to Barrow, Cantor was able to show that it is possible to generate, "bigger and bigger infinite sets from ones that we already have. There is no limit to this escalation" he says. "By this means we can create an ever-ascending staircase of infinities," in fact Cantor proved that: "There is no end to this inconceivable infinity of infinities."[7] This was a truly mind blowing revelation. A massive leap in human consciousness.

Because it was so revolutionary, there was a lot of opposition to Cantor's work at the time. It caused a civil war in maths. People didn't want to include it

Figure 14 Georg Cantor.

ARE THERE LIMITS TO REASON?

in the main body of maths because it seemed so bizarre. It actually frightened people because ever since the ancient Greeks everyone had thought there could only be one infinity. Scholars and theologians used it as a proof for the existence of God. It wasn't meant to be something you could actually manipulate!

I want to show you just very briefly, using the simplest example, why it scared them. It won't take long. On one level it's quite comical. We all know that the ordinary whole numbers are made up of 'odd' numbers and 'even' numbers. Here are the first 10.
Odd numbers: 1 3 5 7 9
Even numbers: 2 4 6 8 10
Together they make up the ordinary number line: 1,2,3,4,5,6,7,8,9,10...
So it looks like the odd numbers are exactly half of all the numbers, and the even numbers make up the other half. Which is what most people think. But what happens when you double each number on the number line?

<u>Ordinary number line</u>   <u>Double each number</u>
      1.........(1+1).................=2
      <u>2</u>.........(2+2).................=4
      3.........(3+3).................=6
      <u>4</u>.........(4+4).................=8
      5.........(5+5).................=10
      <u>6</u>.........(6+6).................=12
      7.........(7+7).................=14
      <u>8</u>.........(8+8).................=16
      9.........(9+9).................=18
     <u>10</u>........(10+10)..............=20
.........and so on forever.

You suddenly realise that although it *seemed* as though all the even numbers were only *half* of all the numbers, there are in fact as many even numbers as there are actual numbers! Magic?

This may seem trivial, but remember we are talking about the blueprint from which 'everything' is constructed. Simple though it looks, it is

THE FINAL MYSTERY

telling us something deep about the *real* infinite. Something that is completely unique. Something quite extraordinary to our ordinary world of sense impressions. Infinite collections of things are fundamentally different from finite ones because they can contain 'themselves' within smaller parts of themselves – something that finite things can't do!

Here's another one which shows it even better. It comes from Galileo actually, but he just scratched his head about it. He thought it was merely a curiosity. He didn't build it into a system as Cantor did. What happens if you multiply each number by itself instead of just adding them together? What's called 'squaring' each number.

1x1=1
2x2=4
3x3=9
4x4=16
5x5=25
6x6=36
7x7=49
8x8=64
9x9=81
10x10=100

From this we can see that there are only 10 squares before you reach 100, so they make up only 1/10th of the numbers up to 100, so they must be pretty rare things right? But no, hang on. If you look at the first column of numbers it is just the ordinary number line (here 1 to 10) which can go on and on to infinity. Therefore, every number in the ordinary number line has a square (just x it by itself), so contrary to what we thought, there are just as many squares as there are ordinary numbers. More magic! Hope you are not getting bored. This is 'real' magic!

That's just for single whole numbers. Cantor called these infinities 'countable' infinities, or the smallest infinities. Those that you can match 'one-for-one' with the ordinary number line, 1,2,3,4,5,6,... and so on. Cantor then showed that there are in fact bigger infinities which he called uncountable infinities. These are the decimals, which Cantor proved are infinitely bigger than the ordinary numbers or the fractions. John Barrow says, "this discovery by Cantor – that there are infinities of different sizes and they can be distinguished in a completely unambiguous way – was one of the great discoveries of mathematics."[8] Cantor showed that from any infinite set of things, it was always possible to make an infinitely larger one. But he also demonstrated that there *had* to be what he called an 'Actual Infinite', an infinite that was beyond human ability to understand. And the reason we could never understand it was precisely because it was beyond the power of maths to describe... Beyond the power of maths? For Cantor the Actual Infinite actually existed. Here's how he described it:

The fear of infinity is a form of myopia that
destroys the possibility of seeing the actual infinite, even though
it in its highest form has created and sustains us, and in its secondary
transfinite forms occurs all around us and even inhabits our minds.[9]

No wonder Cantor's contemporaries were wary of him. His life turned out to be rather sad. Most of the heavyweights in the German mathematical establishment turned against him, so that he spent his whole career at a small provincial university. They even tried their best to stop him publishing his work. Unfortunately he also suffered from bouts of mental illness which became steadily more frequent and he died in a university sanatorium in 1918.

David Hilbert (of 'Hilbert Space' fame) later became the leading mathematician of his day, and he championed Cantor's work, referring to it as, 'Cantor's Paradise, from which no one will expel us', which helped to bring it into the mainstream of maths where it has proved to be hugely influential. So important, that Hermann Weyl, one of the greatest of German mathematicians, has described the whole of mathematics as the 'science of the infinite'.

Hocus Pocus

I want to look for a last time at the Standard Model. It is arguably one of the greatest creations of the human mind since science began. Its corner stones are the two theories of Q.E.D. and Q.C.D. Both theories have been proved to be correct by experiment over and over again. In the case of Q.E.D. to an accuracy never before seen in science. Yet the Standard Model has to make use of a trick known as renormalization to get rid of the infinities, as we saw in chapter eighteen. A trick which, even though he discovered it himself, Feynman referred to as 'hocus pocus'. Many physicists thought it was cheating. The same thing as sweeping the infinities under the carpet. Paul Dirac said, "sensible mathematics involves neglecting a quantity when it turns out to be small – not neglecting it just because it is infinitely great and you do not want it!"[10] In his latest book Roger Penrose makes the point that, "the standard model itself is not free of infinities, being merely a 'renormalizable' rather than a finite theory."[11] So it seems that infinities are also somehow integral to the Standard Model. They are part of our reality even though they can be cancelled out very conveniently.

As I have said before, to a physicist the appearance of infinities in the equations is like a bleeping traffic light or a honking siren, telling them that something has gone wrong. You can't predict anything definite if it leads to an infinity. So they are considered great nuisances. For this very reason Weinberg's work was ignored, if you remember, until Gerald t' Hooft (another Nobel winner), found that the infinities could be cancelled. It not only produced a great surge in references to his work, it caused governments to invest millions of dollars in bigger machines to find the predicted particles, and an award from the Nobel committee before the last one was found. It would be hard to win an argument that says infinities don't influence our world. They seem to live on the very 'edge' of the shimmering arrangement of mathematical structure that creates our physical reality.

At the end of the last chapter we saw that the Theory of Everything will only be discovered once we can explain the nature of mathematics. Now, unfortunately, we see that this means infinities as well. When you look at it like this, then things don't seem too promising for finding a Theory of Everything. All the evidence so far seems to suggest, as I said earlier, that we are surrounded by infinities everywhere we look, from the infinite parallel worlds at the microscopic level, to the infinite universes of the multiverse. And because infinities are the biggest things the human mind can comprehend, then perhaps we are obliged to consider an even bigger question. What is it that caused the existence of the infinities? Is the ultimate truth to be forever hidden because we are not able by human intelligence or mathematics to understand the origin of these infinities, in other words the origin of all information? Is this what Cantor meant? If so, we may catch sight of the Holy Grail, but we may never be able to understand it. Is there a limit to human reason? Is this where knowledge ends?

Kurt Gödel

We need to pause a minute to catch our breath before moving forward again. It involves Kurt Gödel, a brilliant and eccentric Austrian American. An absolute hero of mathematics. Although hardly any of the general public have heard of him, he is thought by many to be one of the greatest mathematicians of all time. Freeman Dyson, who occupies Einstein's post as Professor of Physics at the Institute for Advanced Study, says of him, "Gödel was one of the few indubitable geniuses of our century, the only one of our colleagues who walked and talked on equal terms with Einstein"[12]

His prodigious intellect must have given him some charm because he married a pretty blonde 'exotic dancer' called Adele Porkert, whom he met in a night club in Vienna. He enjoyed the café society and belonged to a group of well known philosophers and numerati. So when the Nazis took over Austria he was reluctant to leave as many other academics did. Even after he was beaten up by a bunch of Nazi thugs in the street who thought he was Jewish. The final straw came when the Third Reich found him eligible for military service! By this time war had been declared so he couldn't escape across the Atlantic to the U.S.A., he had to go by the far more difficult route via Siberia, then to Japan (Pearl Harbour hadn't happened yet), and then by ship to San Francisco and across the continent to Princeton, where he took up a post at the Institute for Advanced Study and became great pals with Einstein.

When Murray Gell-Mann (the Nobel laureate who discovered and named 'quarks'), was himself a student there, he used to see them, "walk to work together, and they made a strange-looking couple, like Mutt and

THE FINAL MYSTERY

Jeff. Gödel was so tiny that he made Einstein look quite tall."[13] If you haven't heard of them, 'Mutt and Jeff' were two old cartoon characters with baggy trousers.

There came the time when Gödel had to apply for U.S. citizenship and go before a judge and a board of government officials who would decide if he qualified. The night before he was due to appear, Einstein and another friend thought it might be a good idea to give him a bit of coaching for the next day. They told him, whatever you say don't mention the United States Constitution, because if you criticise the Constitution they will fail you! Right? But since they said to him not to mention the Constitution, of course he spent the whole night thinking about it. When it came to the interview the next morning, despite all the winks and nudges that Einstein could muster, he launched into a detailed explanation of how the Constitution was actually fatally flawed!

Because Einstein was so famous and was his pal, they pretended not to notice and passed Gödel with a gracious nod. Here's another cracker. He had a habit of taking long walks during which he would mutter to himself in German about the things he was thinking. During the war he once took a holiday by the coast, and the locals reported him to the police because they thought he was a German spy trying to make contact with a submarine off-shore!

Towards the end of his life he unfortunately became paranoid and imagined that everyone was trying to poison him, so in the end he died of self-imposed starvation. So it's said. Though I personally prefer to think that he had just forgotten to eat! Rudy Rucker, who is a well known author and professor of mathematics and computer science, met Gödel at the Institute three times in 1972 to discuss his own research. He remembers how Gödel's voice had a sort of singsong quality because he often raised his voice at the end of sentences as though forming a question. Following this he would let his voice trail off into an amused hum accompanied by laughter. As soon as Rucker began to pose a question, Gödel would already anticipate the answer so that a conversation with him seemed like direct telepathic communication.

Incompleteness

So what did Gödel do then? He did something that profoundly affected the whole world of mathematics. But not just maths as we'll see. At the start of the 20th century mathematicians decided that they wanted to link up all the then known theorems of mathematics, just as Euclid had done with geometry at the famous library in Alexandria. They assumed that every maths problem had to have a solution. It would only be a matter of time and effort to find them.

It became a major international effort that started in 1900 at the Second International Congress of Mathematicians held in Paris, and was led by David Hilbert. Everyone, including all the big names in maths, like Bertrand Russell and John Von Neumann, were convinced it must be possible and joined in the effort.

But out of Vienna in 1931 came two theorems, now jointly called 'The Incompleteness Theorem', by a young twenty-something upstart. It deeply shocked the world of mathematics and smashed apart with a single punch the whole international effort. But the most devastating and traumatic consequence was that it actually proved that mathematics could never, in principle, be complete.

It all starts with the ancient Greeks around 2 400 years ago who came up with what is called the Liar Paradox. Over the years since just about every famous philosopher has tried to explain it, but never to everyone's satisfaction. It consists of a single sentence: This statement is false. Although it sounds absurdly simple, it uncovers the very profound fact that there is a limit to human reason!

Let's take a look at it: This statement is false. Now if it is **true** it has to be false. But if it *is* false it can't be true. However, if This statement is false, is **false**, then the statement is true, which in turn would mean that it is actually false, but this would mean that it is true…and so on for ever and ever.

Because Gödel's theorem is so important I am going to use one of the world's foremost mathematicians and computer scientists to describe it. His name is Gregory Chaitin. He has won a Fields Medal. I'm going to change a word that he uses just twice which is a bit troublesome. It's the word 'axiomatic'. Instead, I'm going to substitute 'self-evident'. It has the same meaning. If you get bogged, down just jump to my much shorter definition that comes after his. Here goes:

"Gödel began with the liar paradox, '**This statement is false!**' If it's true, then it's false. If it's false, then it's true. So it can neither be true or false, which is not allowed in mathematics." Or anywhere for that matter. He goes on, "But, Gödel said, let's change things a little. Let's consider '**This statement is unprovable!**'...Well, there are two possibilties. Either this statement is a theorem, is provable, or it isn't provable, it's not a theorem. Let's consider the two cases."

"What if Gödel's statement is provable? Well, since it affirms that it itself is unprovable, then it's false, it does not correspond with reality. So we are proving a false statement, which is very, very bad. In other words, if this statement is provable, then our formal self-evident system is inconsistent, it contains false theorems. That's very, very bad! If we can deduce false results, then our theory is useless. So let's assume this can't happen."

"So by hypothesis, Gödel's statement is unprovable. But that's not so good either, because then it's a true statement (in the sense that it corresponds with reality) that's unprovable. So our formal self-evident theory is incomplete! *Kaput!*"[14]

I imagine this is probably the closest you can get without using maths. You can get the gist of the argument but it's a bit cumbersome. So here is my own much shorter definition taken from Murray Gell-Mann and simplified.

Given any system of established principles in maths, there will always be statements or theorems that are undecidable on the basis of those principles. In other words, there are theorems that cannot, in principle, be shown to be either true or false. This is how Gödel himself described it:

"The human mind is incapable of formulating (or mechanising) all its mathematical intuitions, i.e., if it has succeeded in formulating some of them, this very fact yields new intuitive knowledge, e.g., the consistency of this formalism. This fact may be called the 'incompletability' of mathematics."[15]

The most devastating realisation that comes from the Incompleteness Theorem is simply this: in the broadest and also in the narrowest sense, in any sense you care to mention, it proves that there will always be a truth that lies beyond any mathematical system's ability to prove. A truth outside…a truth beyond maths? But this journey has shown us all along that maths is real. And further, that everything real, everything that 'exists' emerges from mathematics. Yet in the end we find that there will always be a truth beyond mathematics. A truth that escapes definition by any means we know of. A truth which actually exists but can never be proved to exist. Sounds a bit like Cantor's Actual Infinite.

Rucker says, "but if Gödel's Theorem tells us anything, it is this: Man will never know the final secret of the universe."[16] And the reason he says, is because Gödel's theorem proves that, "any system of knowledge about the world is, and must remain, fundamentally incomplete, eternally subject to revision…Reality is, on the deepest level, essentially infinite."[17]

John Barrow says, "any limitations of mathematical reasoning, like those uncovered by Gödel, are thus not merely limitations on our mental categories but intrinsic properties of reality and hence limitations upon any attempt to understand the ultimate nature of the universe."[19]

Are we finally treading the sacred ground of some ultimate truth? If it is beyond mathematics, then we may never be able to understand it. Is this where knowledge ends? It may be a major moment in our story, but it is not yet the end of the journey.

THE FINAL MYSTERY

Prepare to Dive!

Gödel's theorems floored Hilbert's scheme with a single blow. It was that powerful. And since then things have become worse. The first came five years later when Alan Turing a British mathematician and the founder of computer science, published a paper known as the Halting Problem. He theorised what has become known as a Universal Turing Machine which is the same as a modern digital computer but long before they were first constructed. A problem is decidable if it can answer a yes/no question. What Turing found is that there are many problems which are undecidable by either man or machine. The Halting Problem is known to be unsolvable. What he demonstrated is that there are limits to computation. In other words, there does exist an actual limit to what can ever be known. It backs up what Cantor and Gödel discovered – that there will always be something that cannot be explained.

Figure 15 Alan Turing.

Turing is better known as a hero of the 2nd World War for his role in breaking the Nazi Enigma Codes at Bletchley Park which was a huge advantage for the Allies. After the war he worked with John von Neumann at the Institute for Advanced Study, and he is credited with being one of the original pioneers of our Information Age. He was also a leading figure in artificial intelligence, and his Turing Test is still considered the bench mark for assessing machine intelligence. He was gay at a time when it was still illegal in Britain and he was arrested for homosexuality. Although it is hard to believe now, he was forced to undergo hormone therapy in an attempt to alter his sexuality. It was all too much for him, and in 1952 he committed suicide by eating an apple laced with cyanide. The bite mark taken out of the Apple company logo is thought by some to be a gesture in honour of the life of Alan Turing.

252

More recently Gregory Chaitin has discovered what is called Chaitin's Constant which he calls Omega. It expresses the probability that a random computer programme will halt, but it turns out to be something definable but not computable. Omega is infinitely complex because it is based on an unsolvable problem – Turing's Halting Problem. But the important point is surely this; because Omega is completely unknowable, it proves yet again that there are limitations to what can ever be known by mathematical reasoning. Chaitin is deeply involved in this mind-bending voyage. He describes his work as a quest to understand the limits of understanding!

It reminds us that, as David Deutsch said earlier, maths itself is infinite. John Barrow agrees. He says that, "mathematics is infinite in extent – there is no end to the number of new structures that might be generated from those that are known."[21] Marcus Chown explains, "Chaitin has shown that there are an infinite number of mathematical facts but, for the most part they are unrelated to each other and impossible to tie together with unifying theorems."[22]

Chaitin has said that mathematics is not so different from physics. Like Penrose, Barrow, Deutsch, Tegmark and the others we've met, he claims that mathematicians stumble upon mathematical facts in the same way as a zoologist in the Amazon jungle might come across a new species of monkey, or a botanist a new plant. This is why he calls maths 'quasi-empirical'. "This is particularly bad news for physicists on a quest for a complete and concise description of the universe" says Chown because, "Chaitin's discovery implies there can never be a reliable 'Theory of Everything', neatly summarising all the basic features of reality in one set of equations."[23] If mathematics is infinite then surely we can go no further? This must be where our knowledge ends.

References.

1. Barrow, John. 2005. *The infinite book*. London: Vintage. p xiv.
2. Maor, E. 1991. *To infinity and beyond*. Princeton: Princeton University Press. p 233.

3. Davies, Paul. 1984. *God and the new physics*. London: Penguin. p 19.
4. Aldworth, Roland. 2001. *Mathematics and the real face of god*. Contemporary Review. Vol. 278. No 1624: pp 283-290.
5. Chown, Marcus. 10th March 2001. *The omega man*. New Scientist. Issue no 2281. p 28.
6. Ibid. p 28.
7. Barrow, John. 1992. *Pi in the sky*. London: Penguin. p 214.
8. Barrow, John. 2005. *The infinite book*. London: Vintage. p 71.
9. Rucker, Rudy. 2005. *Infinity and the mind*. Princeton: Princeton University Press. p 43.
10. Gribbin, John. 1988. *In search of Schrödinger's cat*. London: Corgi. p 259.
11. Penrose, Roger. 2004. *The road to reality*. London: Vintage. p 870.
12. Dyson, Freeman. 1993. *From eros to gaia*. London: Penguin. p 161.
13. Gell-Mann, Murray. 1994. *The quark and the jaguar*. London: Little Brown & co. p 39.
14. Chaitin, Gregory. 2000. *The unknowable*. Singapore: Springer-Verlag. P 12.
15. Wang, Hao. 1974. *From mathematics to philosophy*. New York: Humanities Press. p 324.
16. Rucker, Rudy. 2005. *Infinity and the mind*. Princeton: Princeton University Press. p 158.
17. Ibid. p 161.
18. Barrow, John. D. 1991. *Theories of everything*. London: Vintage. p 184.
19. Ibid., p 183.
20. Chaitin, Gregory. December 2005. *Omega and why maths has no toes*. +Plus Magazine. http://plus.maths.org/issue37/features/omega
21. Barrow, John. D. 2005. *The infinite book*. London: Vintage. p 255.
22. Chown, M., op. cit. P 28.

Chapter Twenty-Five
The Final Mystery

The Echo of Doom?

There can be no mistake. Physics is facing a crisis. String theory, its most promising road to a Theory of Everything, seems to have hit a dead end. Although trouble had been brewing for years, it came to a head in December 2005 at the 23rd Solvay Conference in Brussels. David Gross (Nobel Prize 2004), a major figure in theoretical physics and chair of the conference, said in his concluding remarks about string theory, "we don't know what we are talking about..." He compared the situation to what it had been at the first Solvay Conference in 1911. Then the best minds were mystified by the current state of affairs. But shortly after, the Quantum Revolution broke upon the world. "They were missing something absolutely fundamental," he said. "We are missing perhaps something as profound as they were back then."[1]

New Scientist ran an editorial on the conference saying, "there is a growing feeling that string theory has run into the sand." It mentioned David Gross's remarks about the need for a leap forward, but added, "though where it will come from is not clear...," because, it went on, "many of the greatest minds in physics were there, but none had an answer."[2] There have been no breakthroughs since.

The most amazing comment came a week later, and it's a blessing that the mass media didn't pick it up at the time. In an interview Leonard Susskind who, as we said earlier, is the father of string theory, admitted that with respect to the horrific 10^{500} solutions of string theory, the 'landscapes' as they are called, "we will be hard pressed to answer the ID critics. One might argue that the hope that a mathematically unique solution will emerge is as faith-based as ID."[3]

I could hardly believe it. The ID he mentioned was – Intelligent Design! People were shocked. But you see his point. When you are faced with what is to all intents and purposes an infinite number of alternatives, where do you go?

You'll be wondering what Steven Weinberg, the Grand Patriarch of theoretical Physics, has to say about this situation? Just a couple of months before, in September 2005, at the same symposium at the temple of Trinity College Cambridge which we talked about earlier, he actually argued for the acceptance of the Anthropic Principle as a legitimate scientific theory. Remember that the Anthropic Principle is the idea that the universe is very precisely fine-tuned to allow for the evolution of our consciousness. If it was constructed any differently we simply would not exist.

He pointed out that until Einstein came along, no one accepted that something as abstract as a 'principle of symmetry' should ever be considered as grounds for a scientific theory. This brought about a revolution in science itself. "Never before had a symmetry principle been taken as a legitimate hypothesis on which to base a physical theory…" But they *were* accepted. They increasingly became the foundation of physical theories, including the Standard Model. So in the light of the 10^{500} landscapes (solutions) to string theory, he pointed out that, "now we may be at a new turning point, a radical change in what we accept as a legitimate foundation for a physical theory." He continued by making the point that the larger the number of possible universes becomes, "the more string theory legitimates anthropic reasoning as a new basis for physical theories." He goes on to say, "…any scientists who study nature must live in a part of the landscape [i.e. a universe] where physical parameters take the values suitable for the appearance of life and its evolution into scientists."[4]

We haven't mentioned Stephen Hawking for a while. In his inaugural speech as Lucasian Professor at Cambridge back in 1979, he endorsed the idea that a Theory of Everything was not far off in time. But more recently in a news report he was quoted as saying: "up to now, most people have implicitly assumed that there is an ultimate theory that we will eventually discover. Indeed, I myself have suggested we might find it quite soon." But he subsequently had doubts. He goes on to say, "maybe it is not possible to formulate the theory of the universe in a finite number of statements."[5]

Though quietly said and scarcely reported on, one of the great names in science has admitted that – it may not be possible to formulate a theory of the universe in a finite number of statements. Ever since human beings first opened their eyes and looked at the world around them, they have been searching for an explanation of how it works. You could even call it the central odyssey of human consciousness. Although I don't want to exaggerate, if Hawking is correct, then maybe we have arrived at an historic moment. Is this where reason ends?

Roger Penrose seems to be equally doubtful. In his latest book he says, "it is a matter of contention whether anything resembling a 'theory of everything' will ever be found."[6]

Even Steven Weinberg appears to accept this. About a Theory of Everything he says: "I don't think that will be possible, because we can already imagine logically consistent laws of nature that don't quite describe the world we see [read M- Theory]. We will always be somewhat disappointed…" He continues, "All human beings, whether religious or not, are caught in a tragic situation of never fully being able to understand the world we are in."[7]

Another Four Billion Years?

Of course there is an amazing amount of other science going on. Huge achievements in biology, neuroscience, medicine, electronics and so on, are happening on an almost daily basis. But the brutal truth is that the search for an explanation of reality is in fact the ultimate science; it is the biggest quest of all. Recall what Rutherford so rudely asserted, that all of science was either physics or stamp collecting. Obviously there are many scientists who would contest this because they value their own areas of research.

THE FINAL MYSTERY

An anecdote helps to put it in perspective though. In the 1980's Steven Weinberg was asked to testify before the U.S. Congress about the need to build the biggest ever particle accelerator known as the Superconducting Super Collider which would have been some fifty miles (eighty kilometres) in circumference. After he had presented his argument he was amused to find that one of the Congressmen said he wondered if the machine would make us find God. Another said if it did, he would support the project. In the end it was cancelled by Congress in 1993.

But that is exactly what the quest is about. It is a search, as Hawking once described it, to discover the mind God. That is why it is the ultimate science – why it represents the central odyssey of our consciousness.

When you think about it, perhaps we shouldn't be surprised that we are up against some deep problems. Consider some of the demons and dragons we've done battle with on this journey. Things like cats that are both alive and dead at the same time. The impossibility of measuring both the position and momentum of a particle at the same time. Bits of atoms existing in an infinite superposition that includes the whole universe! And how about John Bell's Theorem and Aspect and Zeilinger's experiments that seem to say there is no space or time at the atomic level.

Worse still perhaps, we are surrounded by infinite parallel worlds which we can't see, but you can use to do computing. And then our universe is but one of an infinite multiverse. And the greatest mystery of them all – that everything seems to emerge from mathematics which we know is non-physical and exists outside and independent of both time and space. It's pretty obvious that David Gross is right when he says that a profound leap in our understanding is necessary.

Yet we have at least another four *billion* years on this planet if we can finally grow up and get sensible. Human consciousness is only a mere three million years old, at most. But you may be asking: where do we possibly go next? Sorry, it doesn't end just yet. Fasten your seat belt. This is the last ride.

What is Truth?

Considering the strange demons we've come across, particularly in the quantum world, perhaps we should ask ourselves if there is a problem with 'truth' itself? That is what Chris Isham and his colleagues at Imperial College in London are asking. They are exploring things like existence, logic and truth. And because it is so weird, whether quantum theory in particular has a problem with truth. Or is there something wrong with *our* concept of truth? This is quite heady stuff.

Computer programmers regularly use something called Boolean algebra to come up with logical deductions in their programmes. Chris Isham and Jeremy Butterfield (at Oxford University), decided to dig a bit deeper. Robert Matthews describes what happened next. "They found themselves under the foundations of standard mathematics and staring at something far more fundamental. That something is a concept called topos, and it could be the basis of a whole new way of constructing theories of reality."[8]

These topos, or topoi have an extraordinary key feature. Each one gives rise to its own variety of logic, so they could now see Boolean algebra for what it is. Just the most common of many types of algebra, each with its own form of logic.

But guess what? So long as we stick with ordinary classical physics we get the usual true/false answers. "But if we insist on making statements about atoms", Matthews explains, "we must use the logic of quantum topoi and accept the existence of a vast host of realities, all as valid as each other."[9] Sounds familiar doesn't it? We're back to where we started! Perhaps, like M-Theory and Inflation, it is describing the infinite parallel universes that surround us? The interesting thing about exploring topoi, with their different kinds of logic, is that it could possibly lead to new types of thinking. Hopefully ones more successful at explaining the final mystery.

The Future

Because of the big questions that still need answers many things are possible. One of them is that we may be creating reality as we observe it. Earlier I mentioned that this was still considered an option. Recall for a moment the pedigree of Eugene Wigner's CV? The answer then to John Wheeler's question: 'How Come Existence' is that *we* are creating it, but in a way that we don't yet understand.

Frank Tipler, who co-authored with John Barrow the definitive book on the Anthropic Principle, has suggested that the universe is a computer simulation, and that once we find the programme we will inevitably rerun it, which means everyone will live once more! I sincerely hope we will be able to redesign the software enough to prevent war and famine.

This is not as crazy as it sounds when you think about it. If we can already create pretty realistic virtual realities within primitive digital computers now, then almost certainly in a few hundred years time, we will be able to build real ones with the help of quantum computers.

These futuristic suggestions are not so far-fetched when you consider that serious scientists have already proposed creating new universes out of miniature black holes produced in the laboratory! I'm being deadly serious. And big drug companies are already using computer simulated Virtual Physiological Humans to mimic how all our organs work in order to design new drugs. Hopefully we will be able to create realities where avatars, that possess consciousness, are more sensible and there is less suffering. Was this the original idea behind the quantum vacuum? Perhaps according to some new logic we don't yet understand, it was first necessary for our consciousness to evolve in order to create new realities? It has often been suggested that we ourselves are actually avatars in someone else's imperfect game. All the more reason then for us to get it right on this little blue planet before it is our turn.

I'm thinking that human consciousness will become more and more complex as it has been doing since the first tools and the domestication of fire. Perhaps, as many have suggested, we will be able to transfer our consciousness to machines and become immortal so that we can populate the galaxies. We already suspect that galaxies might be connected through quantum entanglement at the microscopic level, so this would give rise to a kind of galactic super internet of consciousness not limited by time and space. Perhaps the dark energy of empty space will play a role in intergalactic communication. These sort of wild speculations are fun to indulge in, but as we discovered earlier, there are also profound limits to human reason. It seems to me, it is unlikely we will ever be able to understand the origin of all information which lies behind the face of the infinite.

But I also believe that this whole fantastic creation, from the dance of superstrings to the infinite multiverse, cannot be meaningless. In fact, the more fantastic it becomes, the more reason to doubt that it has no meaning. When every possible option actually exists, our particular one may have a significance that we cannot yet comprehend.

Evolution and Mathematics

One of the most interesting questions has to be: why has our consciousness evolved to access the world of mathematics – the science of the infinite? Why have we become aware of something non-physical? Something that is outside time and space, something that would exist even if we didn't? Is the reason for our existence hiding somewhere here? Our ability to comprehend the mathematical structure beneath the outward appearance of physical reality is what distinguishes us singularly from all other arrangements of matter. Why has matter come to understand mathematics?

What possible evolutionary advantage does our consciousness of mathematics bestow on us? It is certainly responsible for all the techno-wizardry of the Information Age and our knowledge of how the universe works, but it is quite unnecessary for our survival or reproduction. We would still have dominated the animal kingdom and covered the surface of the Earth much as Homo Erectus did before us. Was it just a mistake that matter evolved to become conscious and capable of reading the blueprint of existence? We can be conscious without knowing

THE FINAL MYSTERY

mathematics. The vast majority of the human race, including myself, manages without! But we couldn't be aware of maths without consciousness. We know pretty convincingly that consciousness comes about naturally when billions of nerve cells are interconnected in a particular way inside our heads. But why mathematics? And even more strangely, access to the infinite? There could hardly be anything more remote from the cut and thrust of survival by natural selection. It suggests that there might be some sort of purpose behind this connection.

Science would say that it was physically inevitable. Once evolution is able to arrange those nerve cells in a certain way, they will automatically bring about 'self-awareness' or consciousness. Perhaps even Neanderthals, who possessed self-awareness, also had a primitive knowledge of number. But gradually, through an interaction with our surroundings, the need to record crops and objects like pots, an understanding of number and how it could be manipulated came about. As our knowledge of maths has increased, so has our understanding of how the physical world works. Remember all the examples we came across in chapter twenty-three? But much deeper than this, as we have discovered on this journey, everything that exists, including ourselves, is a mathematical object. Remember what Barrow said so dramatically – the reason why maths works is because everything *is* at root mathematics. We have been told many times now that maths is 'out there'. We discover it – we don't invent it. And it existed long before the human mind evolved because it is the blueprint from which everything that exists comes into being – from the quantum vacuum to the human brain.

Was it inevitable that the most complex arrangement of matter in the universe would be able to understand the blueprint behind itself and everything else? Surely this has to be significant? Just as inevitable is the fact that a feedback loop is actually taking place here whereby the phenomenon of mathematics, and with it the infinities – are realising themselves through the physical medium of the human mind – there is no 'magic' in between. So it would seem that mathematics and the infinite and humans are somehow intimately connected. But if there is no ulterior motive – no designer we can blame or make responsible, then there is only us.

And hence an even greater - in fact an almighty - obligation laid upon us to behave sensibly and tender the Earth responsibly. Imagine if we are the only intelligent life in this part of the galaxy, perhaps in this whole galaxy. Our evolution from star dust to intelligent consciousness is surely the greatest ever epic. It may not have happened anywhere else.

All that has happened since Time Began

We need to touch down and pause for a minute before taking off again. We need to make contact with the ground, and remind ourselves again that infinities are also functional practical devices – they don't just exist weirdly in the quantum superposition. They are very helpful conveniences used every day of the week by engineers to build skyscrapers and aircraft because they provide the greatest possible accuracy in defining real physical things. I wonder why? Is it because they prescribe what is 'real'? Recall what Feynman and Deutsch said about the infinite parallel paths actually supporting the real path taken by particles.

As I have said a few times now, all the evidence appears to show that the only thing of any significance that has ever happened since time began is that this universe – our 'reality' if you like, has slowly been evolving towards an understanding of itself by growing ever more complex. The vehicle for this momentous event turns out to be the front of the human brain. This is really all that has happened since the universe evolved from the quantum vacuum.

Now we can see that perhaps it goes even deeper. It looks as though the complex arrangement of nerve cells in our heads also represents the evolution towards self-awareness of mathematics and the infinite. Our human consciousness is at the forefront, the very frontier – the actual 'edge' of this astonishing adventure. There *cannot* be any greater destiny. That's why our species with its consciousness is so very important.

It is surely time we grew up and became adult. Time we put away all the suffering from starvation, all the mindless cruelty and violence of wars. Time we took care of the planet and made it a safe place to live. Time we strapped on our armour to embrace this enormous destiny.

Scientists shouldn't really be atheists

This sounds like a contradiction, but is it? Hardly anyone still imagines that God is some considerate sky daddy with a big white beard sitting on a cloud. Scientists, because they are scientists, have an automatic commitment to the laws of cause and effect. Weird though our reality is, it is still based on causality which dictates that there must be a first cause – an 'ultimate explanation' if you like. To deny there is a first cause is the equivalent of saying you believe in perpetual motion machines, or that everything must have been created by magic.

Ever since humanity became conscious and able to communicate, people have been in awe of what caused everything to exist, and they called it God. The current version is the quest to find a 'Theory of Everything', which is what prompted the media to call the latest particle discovered the 'God Particle' (i.e. the Higgs boson). Governments spend huge amounts of taxpayer's money on fundamental science projects like the Large Hadron Collider in order to discover the first cause, which is simply an updated version of God. That is why a scientist claiming to be an atheist is the *real* contradiction.

The only conceivable way you can dispense with a first cause is if everything is infinite. Then there is no beginning and no end to anything. There is no ultimate explanation, no first cause. With all the evidence we have encountered for infinities at the smallest level and infinities at the largest, perhaps this is the way our reality is constructed? The central odyssey of our consciousness has always been the quest to discover what caused everything to exist. It has filled generations with both fear and longing, but now it seems we have arrived at an historic moment. It looks like we will never know.

From the multiverse to M-Theory the evidence seems to suggest that our reality is only one option among an infinite number of mathematically conceivable options. And the mystery deepens further when we realise that the mathematics from which everything is constructed is itself infinite. Yet although mathematics is 'outside' and independent of space and time, it still exists – if it didn't we wouldn't be able to access it. This legitimately allows us to ask: what entirely <u>unimaginable entity</u> is responsible for creating the infinities?

What 'God' looks like

I am inclined to believe that Gödel's theorem is universal, as Barrow has suggested, and that there will always exist a truth which is 'there' but can never be proved. A truth beyond infinity like Cantor's Actual Infinite, but with a supporting cast of Turing's Halting Problem and Chaitin's Omega. This is how I see that final truth. Let's remember again what Freeman Dyson told us about the extent of sheer size:

> Imagine if you can, four things that have very different sizes. First the entire visible universe, second, the planet Earth. Third, the nucleus of an atom. Fourth, a superstring. The step in size from each of these things to the next is roughly the same. The Earth is smaller than the visible universe by about twenty powers of ten. An atomic nucleus is smaller than the Earth by twenty powers of ten. And a superstring is smaller than a nucleus by twenty powers of ten.[11]

What it means is that if you are a super-string then the world of the nucleus of the atom (just the centre), is as big as the universe is to us. Yes, that big! And if you were an atom, then our earth would seem as big as, not just our galaxy. No; the whole of our observable universe. I mean, our little galaxy alone is staggeringly vast. Like eight minutes for light to arrive from our sun. But five years from our nearest star. And the light we see tonight from Andromeda, our nearest galactic neighbour, actually left there before humans existed. And Andromeda is the nearest of hundreds of billions of other galaxies. Yes, billions!

THE FINAL MYSTERY

And please, I urge you – keep imagining it. This is just our 'home' universe. What about the *infinite* universes of Inflation Theory that exist as a multiverse beyond our own? And if it were possible for anything to be more astonishing than this, then please consider that each quantum bit of every atom which exists in all those universes, is itself surrounded by an infinite number of parallel worlds – worlds that you can do computing with. And even then, consider the possible infinite dimensions where all this may be happening.

Finally, keep in mind that the mathematics – the blueprint from which it all emerges, is itself infinite. This is infinity upon infinity upon infinity. And these infinities are real – as real as the quantum computers that exploit them.

Stand outside and look at the stars on a clear night, and just think about the immensity of the infinities that surround us. Infinities both above you, deep within your finger tips and beneath your feet. It should be truly terrifying – yet this is our reality. It is actually *real*!

Although it must be the strangest of all ironies, I am suggesting that science, our most reliable form of knowledge, the only tool we possess to keep us from ignorance, has led us to our first ever understanding of the unimaginable – the face of the unknowable.

I am suggesting that this infinity upon infinity which science has discovered and to which everything leads is perhaps what 'God' looks like. It seems that the origin of all information does exist – but will forever remain unfathomable by humankind.

This is the final mystery.

When once asked if he was religious, Einstein replied, "Yes, you can call it that." He went on, "…try and penetrate with our limited means the secrets of nature and you will find that, behind all the discernible laws and connections, there remains something subtle, intangible and inexplicable. Veneration for this force beyond anything that we can comprehend is my religion. To that extent I am, in fact, religious."[12]

'God' exists – just as the truth of Gödel's theorem exists. Given all that we have discovered, it seems unlikely that there will ever be a Theory of Everything that can be written down.

This does not mean that the scientific enterprise should slow down. On the contrary, it should speed up. One of the single greatest gifts it has given us is the simple realisation that the more we discover – the more there *is* to discover. How scary is that? It sends a tingle down my spine. From believing little planet Earth to be the centre of the universe to the multiverse, each new horizon that is crossed presents us with an even greater horizon. It seems to me, as I said just now, that we are somehow engaged in a drama involving the self-realisation of Cantor's Actual Infinite. The connection between us would seem to be indisputable. There can be no 'magic' in between.

If so, we are an indispensable cog in this process. Our consciousness could just as well have emerged from cabbages rather than two legged creatures. What is important is consciousness itself. It's what gives the multiverse purpose. Recall again for a moment that this whole vast universe in which we find ourselves, had itself to be fine tuned to 119 decimal places just for our consciousness to appear. How frightening in magnitude does that make our destiny? It is surely one of the most shocking truths that science has ever revealed to us.

Science will never be satisfied until we discover why it is that our particular quivering slice of mathematical stability is selected from the endless host of the infinite infinities that surround us. Who knows what exotic species of knowledge are still to be discovered? Things like the 'surreal numbers' that Martin Kruskal of Rutgers University has discovered which dwarf all conventional infinities, or Hugh Woodin's discovery of 'Ultimate L' that describes a multiverse of infinities based on Gödel's work, or Nima Arkani-Hamed's Amplituhedrons which appear to show that space and time actually emerge from geometry. As for the unimaginable entity that exists behind the face of these infinities, I don't think science can take us there.

Conclusion

As I mentioned earlier, if there are other beings with consciousness in this universe or any other, they may be more advanced technologically, but like ours, their consciousness can only extend to what is infinite. It is like an ultimate limit.

So it seems that we are destined never to find a Theory of Everything. Perhaps the Holy Grail of science does not exist, only the endless face of the infinite infinities. I hope you have enjoyed this voyage to the edge of reason. Perhaps we should finish by remembering one last time what it took to get us here.

Recall that if the exact charge on the electron or the precise strength of the nuclear forces had been any different, atoms could not have come into existence. Nor would there be any stars or galaxies, planets or people. Remember also the vast number of universes that are required to obtain just one that is primed for life. And then ask yourself whether human consciousness has no purpose? It seems far more bizarre to assume that it has none. Just like the proverbial 'sound of one hand clapping' or the silence of a tree falling in the forest when there is no one to hear it, this whole fantastic occurrence is meaningless – without consciousness.

All the empirical evidence appears to be telling us that our consciousness is of staggeringly gigantic importance. It is the most important phenomenon ever to have emerged. Given the length of time we have left on this planet (if we can get sensible), then it is abundantly clear that we currently reside within the opening sentence of the epic that will become the story of humanity. It is surely time now to rid ourselves of such primitive notions as nationalism and religious extremism. Time to build an interdependent global community. Time to fulfil our destiny as the leading edge of all that has ever happened.

References

1. News report. 10 December 2005. *Nobel laureate admits string theory is in trouble.* New Scientist. Issue no 2529. p 6.
2. Ibid., p 5.
3. Gefter, Amanda. 17 December 2005. *Is string theory in trouble?* New Scientist. Issue no 2530. p 48.
4. Weinberg, Steven. 2 September 2005. *Living in the multiverse.* Symposium on: Expectations of a Final Theory. Cambridge: Trinity College.
5. News report. 05 April 2005. *The impossible puzzle.* New Scientist. p 34.
6. Penrose, Roger. 2004. *The road to reality.* London: Vintage. p 1028.
7. Weinberg, Steven. 15 September 2008. *There will be less room for religion.* Newsweek. Vol. CL11, No.11. p 51.
8. Matthews, Robert. 14 April 2007. *Impossible things for breakfast.* New Scientist. Issue no 2599. p 32.
9. Ibid, p 33.
10. Dyson, Freeman. 1990. *Infinite in all directions.* London: Penguin. p 18.
11. Walter, Isaacson. 2007. *Einstein his life and universe.* London: Simon & Schuster. p 384.

Epilogue

What are thoughts?

I imagine that most people think of maths as an abstract distant sort of thing that has nothing to do with everyday experience. But every thought that we have, literally everything that we think about or imagine, is just a complex pattern of nerve cells firing. That's all you can see with MRI scans. Perhaps each individual thought has, in some way, a unique structure. We know that every sense impression coming from the world outside our heads is converted to a simple electrical signal that goes to a centre within the brain that decides whether to keep or discard it. The stuff it keeps is shunted to the front of the brain for us to think about, and when we think there are whole networks of nerve cells firing.

But these networks themselves have a mathematical content, just like everything else in the universe, so you have a sort of 'feed-back' loop going on. If everything that exists is mathematics as John Barrow, Max Tegmark, David Deutsch and many others have suggested, then in the human cortex you have a physical object of immense complexity capable of interacting with and comprehending the eternal and timeless logical deductions and connections of mathematics. What it implies is that our thoughts may also be mathematical objects. So maths is not so distant. It would appear to be the very fabric of our thinking. A bit like our brains doing calculus when we see a car approaching before we step out into the road.

In an exciting recent development called neural decoding, scientists have been able to use mathematical patterns in brain activity to predict what photographs people are looking at. First they make a mathematical model of the photos, then they make a mathematical model of the blood flow in the brain while the person looks at the photo using an MRI scan. By comparing the two sets of numbers, a computer programme can match up brain activity with each photo seen by the subjects. It shows that definite patterns exist between photos and brain scans. They were even able to predict what photos the subjects were looking at with high accuracy. It suggests that 'thinking patterns' might be reduced to mathematics.

The well known philosopher A. C. Grayling says, "the great mystery remains, however: how do the intricate, superfast, vastly complex interactions of the brain's billions of neurons through trillions of synapses give rise to mind?" He continues by saying that, "physical occurrences in the brain have physical properties – a position in space, a duration in time, a measurable intensity...", but in contrast he says, "thoughts (about sub-prime mortgages, say, or the political situation in Bulgaria) do not: they do not have a weight, or a colour, or a scent, or any other physical property."[1] Sound familiar? Maths is the only thing we know of that exists, but has no physical properties. Are our thoughts mathematical objects? Until we discover something else with a similar lack of physical attributes perhaps it should remain an option.

It seems as though our awareness of mathematics has nothing to do with the actual process of evolution by natural selection – nothing to do with genes. It seems to have a lot to do with the mechanical and unavoidable fact that when you arrange billions of nerve cells with trillions of synaptic connections in a certain way, they automatically have the complexity, (most particularly in the brains of mathematicians!) to comprehend aspects of the blueprint responsible for everything that exists, including itself – the 'feed-back' loop I was talking about. It's even resonant, if you like, of the way the infinite can contain infinite things within itself. Is that why Newton's mind could figure out, from an apple falling to the ground, exactly how the planets orbit the Sun? It looks as though we are very intimately connected with the mathematics which creates everything, including us. We are inevitably connected to the actual blueprint. After all, there can be no 'magic' in between.

This is how the French mathematician Alain Connes sees it: "When I speak of the independent existence of mathematical reality, I expressly do *not* locate it in physical reality...The mathematician develops a special sense, I think – irreducible to sight, hearing, or touch – that allows him to perceive a reality every bit as constraining as physical reality, but one that's far more stable than physical reality for not being located in space-time."[2] This seems close to the special sense I was talking about. Are mathematicians connecting to the blueprint of reality?

Unlike eternal mathematical structures, physical reality changes all the time. Everything from the clustering of galaxies to the human mind is evolving, but the laws governing how reality behaves do not. If we were ever to discover the mythical Theory of Everything we would still be left with the question: "Where do the laws of nature come from?" The laws themselves must be derived from something.

Where do the laws of nature come from?

We are surrounded by a world of symmetry, from the shape of spiral galaxies and flowers, to snowflakes and human faces. Because they are everywhere they have been studied since ancient times. Here are a few basic examples. If you take a triangle and fold it in half it has changed radically to something half its size, but its triangular shape has been saved despite the change you made. It has what is called bilateral symmetry. If you revolve a four cornered star through 180 degrees you have made a big change, but the star shape looks exactly the same. Again something has been conserved. The star has what is called rotational symmetry. If you look in a mirror you are seeing reflectional symmetry. You look the same but your image is reversed.

Surprisingly, symmetry is at the very heart of what creates our reality. Until Einstein came up with Special Relativity, everyone thought that symmetry was just the way the laws of nature worked. In other words, that it was the laws of nature that created the symmetry we see. Einstein's revolution, as Weinberg pointed out (chapter 25), was to discover that it was the symmetry principles which created the actual laws and not the other way around.

David Gross, (Nobel Prize in Physics, 2004) explains. "Einstein's great advance in 1905 was to put symmetry first, to regard the symmetry principle as the primary feature of nature that constrains the allowable dynamical laws."[3]

EPILOGUE

In *Symmetry and the beautiful universe* (2004), Leon Lederman (Nobel Prize in physics, 1988) confirms this priority. "Fundamental symmetry principles dictate the basic laws of physics, control the structure of matter, and define the fundamental forces in nature." It could hardly be more clear.

Since Einstein's discovery every major advance has been made using symmetry principles. Concrete examples have been the use of symmetry principles to predict the existence of new particles like quarks and the W and Z particles. In fact, the whole Standard Model – the most accurate theory in the history of science – is based on principles of symmetry. The most recent example is the discovery of the Higgs boson at CERN.

But a symmetry principle is not a 'thing' or a 'force', it is an uncontaminated abstract mathematical structure of pure information. In the same paper for The Proceedings of the National Academy of Sciences, David Gross goes on to say: "Today we realise that symmetry principles are even more powerful – they dictate the form of the laws of nature."[4]

In other words, they dictate the structure of physical reality, and hence the evolution of our consciousness from star dust.

So the answer to the question: where do the laws of nature come from, is that they are derived from principles of symmetry which are independent of space and time. You might legitimately ask: so where do the principles of symmetry come from? They can only arise from (assuming no magic in between), the infinite world of mathematical options.

Are symmetry principles then created by us inside our heads – I don't think so! This is the final proof that mathematics is not created by the human mind. It is the *only* phenomenon that we know of which is independent of time and space and from which our reality derives – let the Revolution begin!

Can science connect with religion?

But if mathematics is real, then so is the infinite. This raises a surprising question: Are we then the physical three-dimensional offspring, if you like, of the infinite? Is that why Cantor and Gödel were able to understand it?

No wonder then that, at the edges of our particularly limited dimensional configuration we appear to bump into the infinite in all directions. Is this why there isn't a simple solution? Is this why reality is so fantastic? Is it because everything is constructed from the infinite? It is as though our normal physical reality is actually surrounded by, you might even say 'cradled', within the arms of the infinite. Don't worry, I haven't suddenly had a serious wobble. Remember Feynman and Deutsch (and others) who explained how it is that the infinite parallel worlds of the superposition (now used by quantum computers), actually support and make the 'real' path of atomic particles possible. That's all I meant. But it might also be the conceptual foundation of a religious point of view.

At the heart of the matter there appears to be something else quite remarkable. If numbers describe our world and they are infinite, then it follows that they can describe every possible world – including the world of thoughts of an alien in another universe. This suggests a somewhat god-like capability.

It seems to reinforce the rather numbing fact that there cannot be anything other than the infinite except, of course, whatever created the infinite. What Einstein called 'this force beyond anything that we can comprehend'; the first cause – Cantor's Actual Infinite – that which is beyond the power of mathematics to describe, because it is the origin of both mathematics and the infinite. Is this why Weinberg and Hawking and Penrose now have doubts? Is this why Hawking believes that it may not be possible to describe the universe within a finite number of statements?

There is something else that is perhaps worth a little thought though. It goes like this: it appears that our consciousness – that which has made the universe aware of itself – must have evolved from the first cause – what I have called the 'unimaginable'. There is no room for anything else. As I keep saying, there is no 'magic' in between. So whether we like it or not, we are inevitably part of its specification. From the no-beginning and no-ending it was pregnant with the possibility of the quantum vacuum and us and this present moment. You could even say that our consciousness is inevitably made in its image, if you like. I don't, of course, mean that there is any resemblance between us, but I do mean that our consciousness of this unimaginable entity actually connects us in a direct causal and empirical link. We are ultimately part of it. Supporting evidence, if you like, comes from the quantum superposition that includes the whole universe, and John Bell's Theorem and the experiments with entanglement. If everything is connected in this way, then our consciousness has to be connected to this First Cause. It is what created, gave birth, if you like, to this whole fantastic occurrence from the infinite infinities and mathematics, to the vacuum and the multiverse, to our universe and us. You and I are both thinking about it right now. That is itself a connection.

As I said earlier, I don't think our thoughts are pure 'nothing'. They have a physical and therefore mathematical basis. Most people have a fixed mindset that convinces them that there can be no connection between mathematics and things like human relationships, or thoughts or aspirations. But things like goodness, love, empathy and evil, are inevitably attributes that have grown out of the symmetry principles that evolved our reality. And therefore attributes evolved from the first cause. But if these human and personal things are the product of a complex interaction of nerve cells – which ultimately they must be, then they can only be sophisticated patterns of mathematical structures or expressions that are vastly more complex than we can currently imagine. So it might be said that we are not only connected to and part of the first cause, it is within everything, from the virtual particles popping in and out of the quantum vacuum to the firing neurones that make up the thoughts in your head. It is closer to you than yourself.

Although it might appear that our consciousness arose by chance, it must have always existed as one option among the many mathematically conceivable options available. One might be tempted to say – as one possibility in the mind of a designer. But that would be quite wrong because it sounds too supernatural. There is nothing supernatural about it. Scientific experiment has shown us that we are surrounded by every possible outcome each jiffy (Planck moment) of our lives (think superposition and quantum computers). We are an integral part, a mechanical artefact if you like, of the first cause – the unimaginable. It is therefore closer to us than our own heartbeat. It seems a bit premature, based purely on a limited 3-dimensional + 1 time observation, to insist that such a first cause does not exist. It is the only possible mechanism that can contrive the quantum vacuum as an option, as well as us.

It is perhaps the reason we are here

What is Information?

For most of us information is not much use if it doesn't carry meaning. But what is meaning? Take the example of the words god and satan. The word god has meaning, so does dog - But 'gdo' has none. In the same way if you change the order of the letters in satan you get santa, two very different meanings. How about art, tar and rat!

The simple swapping around of letters or symbols makes a huge difference because of *meaning*, and yet meaning doesn't appear to be something you can measure – yet it exists. You can hardly say it doesn't exist, if it didn't, we wouldn't know anything. This is exciting because if meaning is something abstract then it does not take up space. So it can't be physical, so what is it?

Paul Davies says, "an alternative view is gaining in popularity: a view in which *information* is regarded as the primary entity from which physical reality is built... Importantly, it is not merely a technical change in perspective, but represents a radical shift in world view."[5]

Orthodox Information Theory dates from 1948 when Claude Shannon at Bell labs published *A Mathematical Theory of Communication*. But

EPILOGUE

Information Theory is not actually about meaning, it's about the *transmission* of information. The content or meaning of a message is irrelevant to its transmission. It restricts itself to the limits of data compression and communication which reduces content to bits of 1's and 0's or 'on' and 'off'. It took a while for information theory to take off, but after Intel produced its first microprocessor in 1971 the information age exploded, and now literally everything physical can be reduced to information, but it says nothing about *meaning* which is abstract.

There is a whole world of the abstract out there. Take protein folding. Proteins are made up of just twenty amino acids and like the letters of the alphabet the order in which they are sequenced is very important to their function. Many diseases are the result of wrong sequences. But the number of possible combinations of amino acids making up proteins is greater than 10^{130}. That is vastly more than the total number of atoms in the universe. But all those possible combinations do exist, but where do they exist?

Similarly, there are more ways of arranging a deck of 52 cards (known as 52 shriek!), than there have been seconds since the universe began. But the amazing thing is that each of those arrangements must exist – they contain information so they must possess the quality of existence. But where do they exist? How about probability? It can be measured but is it physical? In the 1950's Shannon worked out that there are more than 10^{120} possible chess games. These chess moves must exist, but they can't possibly be located within space – they are far more numerous than the atoms in the universe. So where are they located? And like Bertrand Russell maintained if you can think something it must exist.

And then there is our main theme – the abstract world that is described by mathematical symbols. A world which exists but is independent of humans and is responsible for our physical reality. This world contains information. It *is* information in its purest form. The symmetry principles that gave rise to the laws of Nature are fundamental informational constructs. So perhaps information is primary as Davies and others are now suggesting. But then I am reminded once again of Gödel. We can never discover the origin of information because that origin will also have to be information, which will then itself require explanation by means of more information. And so on... Will we ever be able to explain everything?

References.

1. Grayling, A. C. *The marvellous mystery of mind*. New Scientist. No 2676: page 50.
2. Changeux, Jean-Pierre and Connes, Alain. Conversations on mind, matter, and mathematics. (Translated DeBevoise M. B.) 1995. Princeton. Princeton University Press. P 28.
3. Gross, David. *The role of symmetry in fundamental physics*. 1996. Proceedings of the National Academy of Sciences. Vol. 93 no. 25
4. Ibid.
5. Davies, P. & Gregersen, N.H. (2010). *Information and the nature of reality*. Cambridge: Cambridge University Press. P 75.

Images and Illustrations

Figure 1, page 5 (adapted). Evolution of the universe.
NASA. (March 16 2006). Ringside seat to the universe's first split second. [Illustration]. Retrieved September 19, 2013 from: http://www.nasa.gov/images/content/144789main_CMB_Timeline75_lg.jpg

Figure 2, page 13 (adapted). Evolution of life.
BBC. n.d. *History of life on earth.* [Illustration].
Retrieved September 23, 2013, from:
http://static.bbci.co.uk/naturelibrary/3.1.1.4/images/timeline1.jpg

Figure 3, page 15. Charles Darwin.
Henry Maull & John Fox. (1854). Charles Darwin. [photograph].
Retrieved September 18, 2013 from:
http://upload.wikimedia.org/wikipedia/commons/thumb/2/2e/Charles_Darwin_seated_crop.jpg/455px-

Figure 4, page 34. The Goddess Ishtar.
BableStone (photographer) n.d. Goddess Ishtar queen of the night. [baked clay relief]. Retrieved September 19, 2013 from:
http://upload.wikimedia.org/wikipedia/commons/thumb/2/22/British_Museum_Queen_of_the_Night.

Figure 5, page 42. Egyptian entertainment.
Tomb of Nebamun. (ca. 1400 – 1350 BCE). Musicians and dancers. [wall fresco, British Museum]. Retrieved September 20, 2013 from:
http://schools.nashua.edu/myclass/lavalleev/Art%20History%20Pictures/ch03/3-31.jpg

Figure 6, page 72. Albert Einstein.
[Untitled photograph of Albert Einstein]. Retrieved April 6, 2013 from:
http://www.sciencekids.co.nz/pictures/scientists.html

Figure 7, page 80. Gravity near a massive body.
Norton, John D. (illustrator). n.d. The geometry of space. [computer image]. Retrieved September 20, 2013 from:
http://www.pitt.edu/~jdnorton/teaching/HPS_0410/chapters/general_relativity_massive/rubber_sheet_1.jpg

Figure 8, page 93. Horn antenna at Bell Labs.
Bell Labs. (1964 – 1965). Horn antenna, Holmdel, New Jersey. [photograph]. Retrieved September 20, 2013, from:
http://www.cr.nps.gov/history/online_books/butowsky5/images/astro4k1.jpg

Figure 9, page 132. Young's double slit experiment
Deporres, Daphne. (08.11.2010), So are you a wave or a particle? [diagram]. Retrieved September 21, 2013, from:
http://neuralentanglement.com/wp-content/uploads/2010/11/Double-slit-300x143.png

Figure 10, page 159. Paul Dirac.
[Untitled photograph of Paul Dirac]. Retrieved April 6, 2013 from:
http://www.sciencekids.co.nz/pictures/scientists.html

Figure 11, page 195. Chaotic Inflation.
Bluegrass Pundit. (August 4, 2011). Do we live inside a bubble? [Illustration of chaotic inflation]. Retrieved 9 April, 2013, from:
http://scinewsblog.blogspot.com/2011/08/scientists-test-multiverse-theory.html

Figure 12, page 215. Aspect experiment.
Stamatiou, George. (22 October, 2008). Single-channel Bell Test. [diagram]. Retrieved September 21, 2013, from:
http://en.wikipedia.org/wiki/File:Single-channel_Bell_test.png

IMAGES AND ILLUSTRATIONS

Figure 13, page 233. The Mandelbröt Set.
Wikipedia: Featured Pictures. File: Mandel zoom 03 seehorse.jpg [computer image]. Retrieved April 7, 2013, from:
http://en.wikipedia.org/wiki/File:Mandel_zoom_03_seehorse.jpg

Figure 14, page 242. Georg Cantor.
Wiki Media. Georg Cantor [photograph]. Retrieved April 7, 2013, from:
http://en.wikipedia.org/wiki/File:Georg_Cantor2.jpg

Figure 15, page 252. Alan Turing.
[Untitled photograph of Alan Turing]. Retrieved April 9, 2013, from:
http://www.cctvcamerapros.com/Alan-Turing-Computer-Sciences/369.htm

Bibliography

Aczel, A. D. (1996). *Fermat's last theorem*. New York: Four Walls Eight Windows.
Aczel, A.D. (2000). *The mystery of the aleph*. London: W.S.P.
Aczel, A. D. (2002). *Entanglement. The greatest mystery in physics*. Chichester, England: John Wiley & sons, Ltd.
Adair, R. K. (1989). *The great design*. Oxford: Oxford University Press.
Appleyard, B. (1992). *Understanding the present*. London: Picador.
Ardrey, R. (1967). *African genesis*. London: Fontana Books.
Arnold, A. (1992). *The corrupted sciences*. London: Paladin.
Atkins, P. (1994). *Creation revisited*. London: Penguin Books.
Atkins, P. (2003). *Galileo's finger*. Oxford: Oxford University Press.
Atkins, P. (2011). *On Being*. Oxford: Oxford University Press.
Balaguer, M. (1998). *Platonism & Anti-Platonism in Mathematics*. Oxford: Oxford University Press.
Barbour, J. (1999). *The end of time*. London: Weidenfeld & Nicholson.
Barraclough, G., & Overy, R. (Eds.). (1993). *The Times history of the world* (4th ed.). London: Times Books.
Barrow, J. D. (1990). *The world within the world*. Oxford: Oxford University Press.
Barrow, J. D. (1992). *Pi in the sky*. London: Penguin Books.
Barrow, J. D. (1992). *Theories of everything*. London: Vintage.
Barrow, J. D. (1998). *Impossibility the limits of science and the science of limits*. Oxford: Oxford University Press.
Barrow, J. D. (1999). *Between inner space and outer space*. Oxford: Oxford University Press.
Barrow, J. D. (2000). *The book of nothing*. London: Johnathan Cape.
Barrow, J.D. (2005). *The infinite book*. London: Vintage.
Barrow, J. D., Davies, P. C., & Harper, Jr., C. L. (Eds.). (2004). *Science and ultimate reality*. Cambridge: Cambridge University Press.
Barrow, J. D., & Silk, J. (1985). *The left hand of creation*. London: Unwin Paperbacks.
Barrow, J. D., & Tipler, F. J. (1989). *The anthropic cosmological principle*. Oxford: Oxford University Press.

Barrow, J. D. (2011). *The Book of Universes*. London: Bodley Head.
Barry, R. (1996). *A theory of almost everything*. Oxford: One World.
Bell, J. S. (1988). *Speakable and unspeakable in quantum mechanics*. Cambridge: Cambridge University Press.
Bohm, D. (1983). *Wholeness and the implicate order*. London: Ark Paperbacks.
Bohm, D. (1987). *Unfolding meaning*. London: Ark Paperbacks.
Bohm, D., & Hiley, B. J. (1995). *The undivided universe*. London: Routledge.
Bronowski, J. (1966). *The identity of man*. London: Heineman.
Bronowski, J. (1973). *The ascent of man*. London: BBC.
Carey, J. (Ed.). (1996). *The Faber book of science*. London: Faber and Faber.
Carrigan.Jr., R. A., & Trower, W. P. (Eds.). (1990). *Particles and forces at the heart of matter*. New York: W. H. Freeman and company.
Casti, J. L. (1989). *Paradigms lost*. London: Cardinal.
Casti, J. L. (1993). *Searching for certainty*. London: Abacus.
Chaitin, G. J. (1999). *The unknowable*. Singapore: Springer-Verlag.
Changeux, J.P., & Connes, A. (1995). *Conversations on mind, matter, and mathematics*. (Translated by DeBevoise, M.B.). Princeton: Princeton University Press.
Chown, M. (1993). *Afterglow of creation*. London: Arrow Books.
Chown, M. (2001). *The universe next door*. London: Headline Book Publishing.
Clegg, B. (2006). *The God Effect*. New York: St. Martin's Press.
Clegg, B. (2007). *Infinity*. New York: Carroll & Graf Publishers.
Close, F. (2011). *The Infinity Puzzle*. Oxford: Oxford University Press.
Cornell, J. (Ed.). (1991). *Bubbles, voids and bumps in time: the new cosmology*. Cambridge: Cambridge University Press.
Coveney, P., & Highfield, R. (1991). *The arrow of time*. London: Flamingo.
Cox, B. & Forshaw, J. (2011). *The quantum universe: everything that can happen does happen*. London: Allen Lane.
Davies, P. (1985). *Superforce*. London: Unwin Paperbacks.
Davies, P. (1989). *The cosmic blueprint*. London: Unwin.
Davies, P. (Ed.). (1989). *The new physics*. Cambridge: Cambridge University Press.

Davies, P. (1990). *God and the new physics*. London: Penguin Books.
Davies, P. (1990). *Other worlds*. London: Penguin Books.
Davies, P. (1992). *The mind of God*. London: Simon & Schuster.
Davies, P. (1994). *The edge of infinity*. London: Penguin Books.
Davies, P. (1995). *About time*. London: Penguin Books.
Davies, P. (1995). *About time*. London: Viking.
Davies, P. (2006). *The goldilocks enigma*. London: Allen Lane.
Davies, P. C. (1982). *The accidental universe*. Cambridge: Cambridge University Press.
Davies, P. C. (1986). *The forces of nature* (2nd ed.). Cambridge: Cambridge University Press.
Davies, P. C., & Brown, J. (Eds.). (1988). *Superstrings a theory of everything?* Cambridge: Cambridge University Press.
Davies, P. C., & Brown, J. R. (Eds.). (1986). *The ghost in the atom*. Cambridge: Cambridge University Press.
Davies, P. & Gregersen, N. H. (2010). *Information and the nature of reality*. Cambridge: Cambridge University Press.
Davis, P. J., & Hersh, R. (1990). *The mathematical experience*. London: Penguin Books.
Dawkins, R. (2006). *The God delusion*. London: Bantam Press.
D'Espagnat, B. (1989). *Reality and the physicist*. Cambridge: Cambridge University Press.
Deutsch, D. (1998). *The fabric of reality*. London: Pengiun Books.
Deutsch, D. (2011). *The beginning of infinity*. London: Allen Lane.
Doxiadis, A. (2000). *Uncle Petros and Goldbach's conjecture*. London: Faber and Faber.
Drexler, K. E., Peterson, C., & Pergamit, G. (1992). *Unbounding the future*. London: Simon & Schuster.
du Sautoy, M. (2003). *The music of the primes*. London: Fourth Estate.
Dyson, F. (1990). *Infinite in all directions*. London: Penguin Books.
Dyson, F. (1993). *From eros to gaia*. London: Penguin Books.
Ferguson, K. (1994). *The fire in the equations*. London: Bantam Press.
Ferris, T. (1990). *Coming of age in the milky way*. London: Vintage.
Ferris, T. (Ed.). (1991). *The world treasury of physics, astronomy and mathematics*. London: Little, Brown and Company.
Feynman, R. P. (1986). *Surely you're joking mr. Feynman*. London: Unwin Paperbacks.

Feynman, R. P. (1990). *Q.E.D. the strange theory of light and matter*. London: Penguin Books.
Feynman, R. P. (1999). *The meaning of it all*. London: Penguin Books.
Feynman, R. P., & Weinberg, S. (1987). *Elementary particles and the laws of physics. 1986 Dirac memorial lectures*. Cambridge: Cambridge University Press.
Fraser, G., Lillestøl, E., & Sellevåg, I. (1994). *The search for infinity*. London: Mitchell Beazley.
Frisch, O. (1991). *What little I remember*. Cambridge: Cambridge University Press.
Gamow, G. (1985). *Thirty years that shook physics*. New York: Dover Publications, Inc.
Gell-Mann, M. (1994). *The quark and the jaguar*. London: Little, Brown and Company.
Gleick, J. (1992). *Genius. Richard Feynman and modern physics*. London: Little, Brown and Company.
Goswami, A., Reed, R. E., & Goswami, M. (1993). *The self-aware universe*. London: Simon & Schuster.
Greene, B. (1999). *The elegant universe*. London: Johnathan Cape.
Greene, B. (2004). *The fabric of the cosmos*. London: Penguin Books.
Greene, B. (2011). *The hidden reality*. London: Allen Lane.
Gregory, B. (1988). *Inventing reality*. New York: John Wiley & Sons, Inc.
Gribbin, J. (1988). *In search of Schrödinger's cat*. London: Corgi Books.
Gribbin, J. (1988). *In search of the big bang*. London: Corgi Books.
Gribbin, J. (1988). *The omega point*. London: Corgi Books.
Gribbin, J. (1993). *In the beginning*. London: Viking.
Gribbin, J. (1995). *Schrödinger's kittens and the search for reality*. London: Weidenfeld & Nicholson.
Gribbin, J. (1998). *In search of susy*. London: Penguin Books.
Gribbin, J., & Rees, M. (1991). *Cosmic coincidences*. London: Black Swan.
Gribbin, J. (2009). *In search of the multiverse*. London: Penguin Books.
Gribbin, J. (2011). *Alone in the universe*. Hoboken, New Jersey: John Wiley & Sons.
Hawking, S. (1988). *A brief history of time*. London: Bantam Press.

Hawking, S. (1996). *The illustrated a brief history of time*. London: Bantam Press.
Hawking, S. (2001). *The universe in a nutshell*. London: Bantam Press.
Hawking, S., & Penrose, R. (2000). *The nature of space and time*. Woodstock, Oxfordshire: Princeton university Press.
Hawking, S, & Mlodinow, L. (2010). *The grand design*. London: Bantam Press.
Hayes, C. J. H., Baldwin, M. W., & Cole, C. W. (1968). *History of western civilization* (2nd ed.). London: Collier-Macmillan Limited.
Hayes, M. (1994). *The infinite harmony*. London: Weidenfeld & Nicolson.
Heisenberg, W. (1963). *Physics and philosophy. The revolution in modern science*. London: George Allen & Unwin Ltd.
Heisenberg, W. (1970). *Natural law and the structure of matter*. London: Rebel Press.
Hey, T., & Walters, P. (1987). *The quantum universe*. Cambridge: Cambridge University Press.
Hilgevoored, J. (Ed.). (1994). *Physics and our view of the world*. Cambridge: Cambridge University Press.
Hofstadter, D. (1986). *Metamagical Themas*. London: Penguin Books.
Hofstadter, D. R. (2000). *Gödel, Escher, Bach: An eternal golden braid*. London: Penguin Books.
Holt, J. (2012). *Why does the world exist?* London: Profile Books.
Howard, M., & Louis, W. R. (Eds.). (1998). *The Oxford history of the twentieth century*. Oxford: Oxford University Press.
Ifrah, G. (2000). *The universal history of numbers 1. the world's first number-systems*. London: The Harvill Press.
Isaacson, W. (2007). *Einstein his life and universe*. London: Simon & Schuster.
Kafatos, M., & Nadeau, R. (1990). *The conscious universe*. New York: Springer-Verlag.
Kaku, M. (1995). *Hyperspace*. Oxford: Oxford University Press.
Kaku, M. (1998). *Visions*. Oxford: Oxford University Press.
Kaku, M. (2011). *Physics of the future*. London: Penguin Books.
Kanigel, R. (1992). *The man who knew infinity. A life of the genius Ramanujan*. London: Abacus.

Kaplan, R. (1999). *The nothing that is*. London: BCA.
Krauss, L. (1990). *The fifth essence*. London: Vintage.
Krauss, L. (2012). *A universe from nothing*. New York: Free Press.
Lavine, S. (1994). *Understanding the infinite*. Cambridge, Massachusetts: Harvard University Press.
Layzer, D. (1990). *Cosmogenesis*. Oxford: Oxford University Press.
Lederman, L. M., & Schramm, D. N. (1989). *From quarks to the cosmos*. New York: Scientific American Library.
Lindley, D. (1993). *The end of physics*. New York: BasicBooks.
Maor, E. (1991). *To infinity and beyond*. Princeton: Princeton University Press.
Maddox, J. (1998). *What remains to be discovered*. London: Macmillan.
Matthews, R. (1992). *Unravelling the mind of god*. London: Virgin.
Morris, R. (1992). *The edges of science*. London: Fourth Estate.
Murdoch, D. (1989). *Niels Bohr's philosophy of physics*. Cambridge: Cambridge University Press.
Ne'Eman, Y., & Kirsh, Y. (1986). *The particle hunters*. Cambridge: Cambridge University Press.
Overbye, D. (1993). *Lonely hearts of the cosmos*. London: Picador.
Pagels, H. (1992). *Perfect symmetry. The search for the beginning of time*. London: Penguin Books.
Pagels, H. R. (1994). *The cosmic code*. London: Penguin Books.
Pais, A. (1983). *Subtle is the Lord the science and life of Albert Einstein*. Oxford: Oxford University Press.
Peacock, R. E. (1989). *A brief history of eternity*. Eastbourne, Sussex: Monarch.
Peat, D. F. (1992). *Superstrings and the search for a theory of everything*. London: Abacus.
Penrose, R. (1990). *The emperor's new mind*. London: Vintage.
Penrose, R. (1994). *Shadows of the mind*. Oxford: Oxford University Press.
Penrose, R. (2005). *The road to reality*. London: Vintage Books.
Penrose, R. (2010). *Cycles of time*. London: Bodley Head.
Polkinghorne, J. (1998). *Belief in God in an age of science*. New Haven and London: Yale University Press.
Polkinghorne, J. C. (1986). *The quantum world*. London: Penguin Books.

BIBLIOGRAPHY

Rae, A. (1986). *Quantum physics: illusion or reality*. Cambridge: Cambridge University Press.
Randall, L. (2006). *Warped passages*. London: Penguin Books.
Randall, L. (2011). *Knocking on heaven's door*. London: Bodley Head.
Reader, J. (1986) *The rise of life*. London: William Collins.
Rees, M. (2000). *Just six numbers*. London: Weidenfeld & Nicholson.
Rees, M. (2002). *Before the beginning*. London: Free Press.
Rees, M. (2002). *Our cosmic habitat*. London: Weidenfeld & Nicholson.
Riordan, M., & Schramm, D. (1993). *The shadows of creation*. Oxford: Oxford University Press.
Roberts, J. M. (1993). *History of the world*. London: BCA.
Rohrlich, F. (1989). *From paradox to reality*. Cambridge: Cambridge University Press.
Rucker, R. (2005). *Infinity and the mind.* Oxford: Princeton University Press.
Ruelle, D. (1993). *Chance and chaos*. London: Penguin Books.
Russell, B. (1969). *History of western philosophy* (2nd ed.). London: George Allen &Unwin Ltd. (Original work published 1946)
Sagan, C. (1981). *Cosmos*. London: Macdonald & Co (Publishers) Ltd.
Schrödinger, E. (1967). *What is life & mind and matter*. Cambridge: Cambridge University Press.
Singh, S. (1997). *Fermat's last theorem*. London: Fourth Estate.
Smolin, L., & Smolin, L. (2000). *Three roads to quantum gravity*. London: Weidenfeld & Nicholson.
Smolin, L. (2006). *The trouble with physics*. London: Penguin Books.
Smoot, G. (1995). *Wrinkles in time*. London: Abacus.
Squires, E. (1986). *The mystery of the quantum world*. Bristol, England: Adam Hilger Ltd.
Standage, T. (2000). *The Neptune file*. London: Allen Lane. The Penguin Press.
Stewart, I. (1990). *Does god play dice*. London: Penguin Books.
Stewart, I. (1996). *From here to infinity*. Oxford:)xford University Press.
Stewart, I., & Golubitsky, M. (1993). *Fearful symmetry. Is God a geometer?* London: Penguin Books.
Taylor, J., Professor. (1993). *When the clock struck zero*. London : Picador.

Teilhard de Chardin, P. (1969). *The phenomenon of man*. London: Fontana Books.
Tipler, F. (1996). *The physics of immortality*. London: Pan Books.
Weinberg, S. (1981). *The first three minutes*. London: Flamingo Fontana Paperbacks.
Weinberg, S. (1993). *The discovery of subatomic particles*. London: Penguin Books.
Weinberg, S. (1993). *Dreams of a final theory*. London: Vintage.
Weinberg, S. (2003). *Facing up*. London: Harvard University Press.
Weinberg, S. (2009). *Lake views*. London: Harvard University Press.
White, M., & Gribbin, J. (1992). *Stephen Hawking a life in science*. London: Penguin Books.
White, M., & Gribbin, J. (1994). *Einstein a life in science*. London: Simon & Schuster.
Wolf, F. A. (1991). *Parallel universes*. London: Paladin.
Young, L. B. (1993). *The unfinished universe*. Oxford: Oxford University Press.
Zeilinger, A. (2010). *Dance of the photons*. New York: Farrar, Straus & Giroux.
Zellini, P. (2004). *A brief history of infinity* (D. Marsh, Trans.). London: Allen Lane. (Original work published 1980)

Index

A

Aczel, Amir, 213, 232
Adair, Robert, 168
Agrarian Revolution, 28
Alexander the Great, 53
Alpher, Ralph, 90, 95, 157
amino acids, 10, 11, 12
Amplituhedrons, 231, 267
Anderson, Carl, 161, 226
Andromeda, 6, 8, 69, 70, 216, 265
Annalen der Physik', 75, 121
Anthropic Principle, 190, 192, 256
Archimedes, 54, 55
Ardrey, Robert, 22
Aristarchos, 54, 57
Aristotle, 53, 54, 55, 66, 108, 221
Arkani-Hamed, Nima, 231, 267
Aspect, Alain, 215
astrobiologists, 11
atavism, 18
atemporal, 215
Atkins, Peter, i, 231
Auschwitz, 130

B

Baade, Walter, 185
Babylonian, 38, 44, 45, 51
Babylonians, 44
Barrow, John, i, 125, 190, 193, 229, 231, 234, 238, 244, 251, 253, 260, 270
Becher, Veronica, 242
Becquerel, Henri, 115, 161
Bell Labs, 92, 280
Bell, John, 137, 148, 212, 213, 214, 216, 258, 275
Bell's Inequality, 137, 214
Bell's Theorem, 214, 217
Bellarmine, Cardinal, 59, 61
Benioff, Paul, 234
Bernoulli, Johann, 65
Bethe, Hans, 157, 165
Big Crunch', 179
Binary Floating-Point Arithmetic, 242

Blackett, Patrick, 118
Bletchley Park, 252
Bohm, David, 149, 212
Bohr, Niels, 88, 122, 131, 144, 211, 214
Boltzmann, Ludwig, 120
Boolean algebra, 259
Brahe, Tycho, 57
Bronowski, 65, 284
Bronowski, Jacob, 65, 130, 135
Bruno, Giordano, 58, 61
bucky balls, 217
Busso, Raphael, 197
Butterfield, Jeremy, 259

C

Cantor, Georg, 242, 281
Casimir Effect, 3, 187
Cavendish Laboratories, 113, 114
Cavendish Laboratory, 111
Cepheid Variables, 68
CERN, 41, 77, 170, 171, 172, 176, 214, 273
Chadwick, James, 118, 158
Chaitin, Gregory, 250, 253
chaotic inflation, 195, 280
Chown, Marcus, 91, 96, 242, 253
Chuang, Isaac, 205, 206
city of Alexandria, 53
classical physics, 119, 120, 137, 259
COBE, 4, 94, 95, 96, 97, 179, 188
Cockroft, John, 118
complementarity, 144
Connes, Alain, 235, 271
Copenhagen Interpretation, 123, 144, 145, 149
Copernicus, Nicolas, 57
cosmological constant', 187
Cramer, John, 139
Cro-Magnon, 25, 26

D

Dalton, John, 108, 109
Dark Ages, 32, 53, 56
Dark Energy, 187
Dart, Raymond, 21

Darwin, Charles, 13, 14, 279
Davies, Paul, 75, 106, 134, 146, 182, 190, 216, 240
Davis and Hersh, 220
de Broglie, Prince Louis Victor, 124
Democritus, 108, 110, 114
Denisovans, 25
Determinism, 129, 131
Deutsch, David, i, 138, 145, 151, 152, 153, 204, 225, 231, 253, 270
Dicke, Robert, 91, 94, 178
Dirac, Paul, 127, 158, 226, 246, 280
DNA, 10, 12, 17, 19, 21, 24, 25, 103
Don Juan, 72
Doppler, Christian, 68
du Sautoy, Marcus, 235
Dyson, Freeman, 105, 106, 139, 168, 191, 212, 225, 238, 247, 265

E

Eddington, Arthur, 82, 84
Egypt, 29, 40, 41, 43, 44, 45, 51, 52, 53, 54, 55, 108
Einstein, 73, 74, 75, 79, 80, 81, 82, 83, 84, 85, 88, 105, 106, 112, 120, 121, 123, 132, 138, 141, 147, 150, 159, 160, 161, 164, 168, 173, 176, 187, 202, 210, 211, 212, 213, 216, 223, 224, 225, 226, 229, 247, 248, 256, 266, 269, 287, 290
Einstein, Albert, iii, 72, 121, 279
electromagnetism, 112, 119, 163, 164, 165, 167, 168, 169, 171, 175
entanglement, 212, 216, 217, 261, 275
Epic of Gilgamesh, 33
Eratosthenes, 55
Espagnat, Bernard D', 215
Euclid, 81, 228, 249
European Centre for Nuclear Research (CERN), 41
Everett, Hugh, 151, 190, 196

F

false vacuum', 181
Faraday, Michael, 110
Fermi, Enrico, 146, 162
Fertile Crescent', 29
Feynman, Richard, 78, 131, 137, 138, 141, 167, 190, 204, 286
Fibonacci sequence, 232
Fifth Postulate, 81
First World War, 73, 83, 118, 120
FitzRoy, Robert, 14
fossils, 13, 19
Franklin, Benjamin, 110

G

G.U.T.s, 176, 180
Galileo, 59, 60, 61, 62, 64, 66, 101, 129, 153, 208, 221, 222, 236, 244, 283
Gamow, George, 88, 157
Gell-Mann, 172, 173, 227, 254, 286
Gell-Mann, Murray, 172, 212, 227, 247, 250
General Relativity, 79, 80, 81, 82, 83, 84, 85, 106, 160, 176, 187
Gilgamesh, 33, 34
Gisin, Nicholas, 216
Glashow, Sheldon, 169
God particle', 176
Gödel, Kurt, 72, 247
graviton, 175
Gribbin, John, 84
Gross, David, 255, 258
Grossman, Marcel, 80, 224
Grover, Lov, 205
Gutenberg, Johannes, 56
Guth, Alan, 5, 178, 180, 191

H

Haldane, J.B.S., 153
Hale, George, Ellery, 69
Halley, Edmund, 64
Halting Problem, 252, 253, 265
Hamilton, William, 159
Hammurabi, 38
Harvard University, v, 17, 68, 164, 228, 236, 288, 290
Hawking, Stephen, iv, 8, 63, 71, 75, 77, 82, 83, 85, 86, 87, 97, 127, 128, 129, 130, 133, 134, 135, 158, 178,

INDEX

179, 182, 184, 196, 203, 209, 234, 237, 256, 257, 258, 274, 286, 287, 290
Heisenberg, Werner, 120
Herschel, William, 66
Hertz, Heinrich, 113
Higgs Field, 169, 176
Higgs, Peter, 169, 176
Hilbert Space', 138, 245
Hilbert, David, 138, 245, 249
Hipparchus, 55
Hippocrates, 108
Hoyle, Sir Fred, 10, 88
Hubble Volume, 193
Hubble, Edwin, 69
Humason, Milton, 70
hydrocarbons, 11

I

Incompleteness Theorem, 249, 251
Inflation, 4, 5, 180, 181, 191, 195, 259, 280
Information Age, 252, 261
Institute for Advanced Study, 105, 150, 211, 225, 247, 252
irrational numbers., 52
Isham, Chris, 259

J

Jammer, Max, 125
jiffy, 104, 106, 208, 240, 276

K

Kant, Emmanuel, 66
Kepler, Johannes, 58
King Alfonso of Spain, 57

L

Laplace, Simon de, 129
Large Hadron Collider, 41, 77, 158, 170, 176, 264
last ice age, 28
Layzer, David, 228
Leavitt, Henrietta, 68

Lederman, Leon, 76
Leibniz, Gottfried, 63
Lemaitre, Georges-Henri, 88
Liar Paradox, 249
Lipershey, Hans, 59
Lucasian Professorship, 63, 64, 158

M

MACHO's, 186, 188
Maddox, John, 176, 183
magnetic monopoles, 180
Mandelbrot Set, 232
Manhattan Project, 146, 147, 149, 150, 164, 167
Maor, Eli, 238
Massachusetts Institute of Technology, 91, 97, 205, 230
Mather, John, 95
Maxwell, James, Clerk, 111
McCarthy era, 151
Measurement Problem', 140
Meer, Simon Van de, 170
Mendeleev, Dimitri, Ivanovich, 109
meson, 165, 169
microwave background, 4, 94, 96, 179, 188
Minoan civilization, 45
Mitochondrial Eve, 24
M-Theory, 105, 160, 196, 259, 265
multiverse, 183, 192, 193, 195, 196, 197, 198, 229, 234, 238, 246, 258, 261, 265, 267, 269, 275, 280, 286
Mycenaeans, 46

N

Nambu, Yoishiro, 196
Napoleon, 130
NASA, 11, 95, 188, 279
Ne'eman, Yuval, 172
neutrinos, 6, 162, 163, 186
Newton, Sir Isaac, 62, 114
Nobel Committee, 171
non-locality', 212, 217
nuclear fusion, 5

O

Oblers' Paradox, 87
obsidian, 29
Origin of Species by Natural Selection, 16

P

Pagels, Heinz, 168
Pauli, Wolfgang, 127, 162, 215
Peebles, Jim, 92
Penrose, Roger, i, 137, 220, 231, 233, 246, 257
Penzias, Arno, 92
Periodic Table, 109, 172
Persians, 44, 45, 47, 48
Phoenicians, 44, 46, 47
Planck length, 104
Planck Scale, 104, 119, 202
Planck time, 104, 106, 240
Planck, Max, 119, 146, 206
Plato, 52, 53, 108, 220, 233
Podolsky, Boris, 212
Polchinski, Joe, 197
Positivists, 145
positron, 161
President Roosevelt, 123, 164
Principia, 64
principle of symmetry', 256
Ptolemy, 55, 57
Pythagoras, 51, 52, 220

Q

Q.C.D, 175, 178, 245
Q.E.D, 168, 169, 175, 178, 225, 245, 286
Quantum Electrodynamics, 168
quantum potential', 149
Quantum Revolution, 115, 119, 129, 131, 143, 202, 210, 211, 225, 235, 255
qubit, 205, 207

R

Red Deer Cave people, 25

Rees, Martin, 183, 192, 197
renormalizable, 246
Riemann, Georg, 81
Room of Mirrors', 208
Rosen, Nathan, 212
Royal Society, 64, 67, 84, 111, 114, 118, 146, 159, 183
Rubin, Vera, 185
Rucker, Rudy, 248
Rutherford, 115, 116, 117, 118, 119, 122, 158, 257
Rutherford, Ernest, 88, 114
Ryle, Sir Martin, 89

S

Salam, Abdus, 169
Schrödinger, Erwin, 126
Schwinger, Julian, 167, 225
Shapley, Harlow, 68
Shor, Peter, 204
SLAC, 171, 173, 180
Slipher, Vesto, 68
Smoky Dragon, 143, 144, 145, 147, 149, 151, 196, 202
Smolin, Lee, 192
Smoot, George, 96, 97
Socrates, 52, 53
Solvay Conference, 88, 127, 211, 255
Sommerfield, Arnold, 81
Special Relativity, 75, 76, 77, 78, 150, 160, 226
Special Theory of Relativity, 79
Standard Model, 167, 175, 176, 189, 202, 245, 256
Stapp, Henry, 214
Sumerians, 33, 35, 56
super string theory, 105
Superconducting Supercollider, 176
supernova, 7, 51, 58, 185, 187, 188
Supernova Cosmology Project, 187, 188
superposition, 136, 138, 140, 141, 143, 146, 196, 202, 203, 205, 207, 216, 217, 258, 263, 274, 275, 276
Susskind, Leonard, 189, 196, 255

INDEX

T

t'Hooft, Gerhardt, 170
Tegmark, Max, i, 195, 230, 270
Thales of Miletus, 51, 110
The central odyssey, 264
The Creation Event, 11
The Multiverse', 191
the scientific method', 55
Theory of Everything, i, 76, 106, 169, 189, 196, 220, 234, 241, 246, 253, 255, 256, 257, 264, 267, 268, 272
Thomson, Joseph, John, 113, 122
Tipler, Frank, 190, 260
Tomanaga, Sin-Itiro, 168
topos, 259
Trinity College, Cambridge, 62, 197, 198
Trojan war, 46
Tryon, Edward, 181
Turing, Alan, 252, 281
Tutankhamon, 42
Twistor theory, 231

U

Uncertainty Principle, 120, 126, 127, 129, 130, 145, 149, 210
Unitarians, 64
Uranus, 66, 67

V

vacuum energy, 188, 189, 190, 191

Von Neumann, John, 147, 213, 249, 252

W

Wallace, Alfred Russell, 15
Walton, Ernest, 118
Weinberg, Steven, iii, 19, 79, 82, 85, 106, 112, 124, 131, 136, 148, 169, 189, 197, 203, 226, 256, 257, 258
Weyl, Hermann, 245
Wheeler, John, i, 81, 143, 151, 190, 260
Wigner, Eugene, 146, 160, 202, 222, 260
Wilson, Robert, 92
WIMP's, 186, 188
WMAP, 4, 188
Wotton, Sir Henry, 60

Y

Young's double slit experiment, 132, 280
Yukawa, Hideki, 164, 227

Z

Zeilinger, Anton, 216
Zweig, George, 172
Zwicky, Fritz, 185

Made in the USA
Columbia, SC
13 August 2018